Train Your Brain

Train Your Brain

A Year's Worth of Puzzles

George Grätzer

Translated by Tom Artin

CRC Press is an imprint of the
Taylor & Francis Group, an **informa** business
AN A K PETERS BOOK

Originally published in Hungarian as *Elmesport egy esztendőre* by Nyitott Könyvműhely Kiadó Bt. in 2008.
Published in German as *Denksport für ein Jahr: 140 mathematische Rätsel aus dem Alltag* by Spektrum Akademischer Verlag in 2010.
Cover illustration by Larisza Pasztricsák.
Interior illustrations by László Réber.

A K Peters/CRC Press
Taylor & Francis Group
6000 Broken Sound Parkway NW, Suite 300
Boca Raton, FL 33487-2742

Printed in the United States of America on acid-free paper
10 9 8 7 6 5 4 3 2 1

International Standard Book Number: 978-1-56881-710-1 (Paperback)

Library of Congress Cataloging-in-Publication Data

Grätzer, George A.
[Elmesport egy esztendőre. English]
Train your brain : a year's worth of puzzles / George Grätzer ; translated by Tom Artin.
p. cm.
ISBN 978-1-56881-710-1 (alk. paper)
1. Word games. 2. Word problems (Mathematics) I. Title.

GV1507.W8G73 2011
793.734--dc22 2010041224

Visit the Taylor & Francis Web site at
http://www.taylorandfrancis.com

and the A K Peters Web site at
http://www.akpeters.com

To Cathy,
for 55 plus years

Table of Contents

Preface xv

The Gym 1

Black Belt 53

Hints 71

Solutions 89

Appendix 221

1. Terminology 221

2. Mathematical Induction 222

3. Some Important Formulas 223

4. Prime Numbers 225

5. Prime Factorization 226

A Friendly Afterword to the Reader 231

Afterword to the English Translation 233

Puzzle Directory

There are three numbers to the right of each puzzle title: the first is the page number of the *puzzle* itself; the second, the page number of the *hint* (if there is one); and the furthest to the right, the page number of the *solution*.

1st Week

A Journey by Ship 1 71 89
Two Games of Chess 1 71 89
A Journey by Air 2 71 90

2nd Week

The Math Whiz .... 2 — 91
The Will .... 3 — 92
The Clock 4 71 92

3rd Week

Nicholas's House .... 4 — 92
An Unusual Year 5 72 93
Competitors, or *The Raise* 5 72 94

4th Week

The Metal Tube .... 5 — 95
The Ancient Greek Copper Lion 6 72 95
The Arab 6 72 96

5th Week

On the Streetcar 7 72 96
Hunting Hares 7 73 96
Telephone Cable—A Long Line? 7 73 97

6th Week

Dinner of Dumplings .... 8 — 97

The Paradox of Protagoras .... 8 — 98
Say How Many Flags Are Flying .... 9 — 99

7th Week
Homework .... 9 — 102
The Superhotel .... 9 — 102
Jealous Husbands .... 10 — 102

8th Week
Another Superhotel 10 73 103
Three-Letter Words 11 73 104
In the Kindergarten 11 73 105

9th Week
How Old Is Susie? 12 73 106
Curious Addition 12 74 106
Exercise in Logic .... 12 — 111

10th Week
The Local Lurches On 13 74 111
Cocoa Island 13 74 112
The Little Troublemaker 14 74 112

11th Week
In the Garden .... 15 — 113
Five Little Decks of Cards 15 74 113
A Clumsy Division 16 75 113

12th Week
Cycle Tour I 16 75 114
More Jealous Husbands .... 17 — 114
Toy Soldiers—On the Double! 17 75 115

13th Week
Big Problem—Chess Problem 18 75 116
Cycle Tour II 18 75 116
With Time and Patience, You'll Go Far .... 19 — 116

14th Week
Pouring Judiciously .... 19 — 119
The Barber 20 75 120
Council of Elders .... 20 — 121

15th Week
The Story of Josephus Flavius .... 21 — 122
Car Trip 21 76 122
The Importance of Plain Speaking 22 76 123

16th Week
The Fly 22 76 125
Astonishing Trick 23 76 125
Prize Question .... 23 — 126

17th Week
Bicycle Motor .... 24 — 128
Each Ninth Drops Out .... 24 — 129
Impressive Card Trick 24 76 129

18th Week
How Did I Know That? 25 77 130
Weight Trick I .... 25 — 131
Knight on the Chess Board 26 77 131

19th Week
Pocket Change .... 26 — 132
Water Pipes .... 26 — 132
Cycle Tour III .... 27 — 133

20th Week
The Cans—Yes, We Can! .... 28 — 134
Two Puzzles in Two Languages 28 77 135
Shopping for Chocolate 29 77 137

21st Week
Instant Addition 29 77 138
Still More Jealous Husbands 30 77 140
Soccer 30 78 141

22nd Week
Prime Number 30 78 142
Cycle Tour IV 31 78 143
Green Cross 31 78 144

23rd Week
Money for Ice Cream 32 78 145
At Lake Michigan I .... 33 — 146
The Envious Cousin 34 78 146

24th Week
A Ship Sails By 35 79 147
At Lake Michigan II .... 35 — 148
Dinner Guests .... 35 — 148

25th Week
Beside/Below .... 36 — 150

Lesson in Economy 37 79 150
Auto Racing 37 79 153

26th Week
Uncle Pops a Quiz 38 79 154
The Nephew Strikes Back .... 38 — 155
The Window Dresser's Error .... 39 — 155

27th Week
At Lake Michigan III .... 39 — 156
Colored Dice 39 79 156
Francesca's Teachers .... 40 — 157

28th Week
Scribble .... 40 — 157
Randolini's Card Trick 41 80 158
Vacation .... 42 — 158

29th Week
A Typical Mathematician .... 42 — 160
Weight Trick II .... 42 — 160
At Lake Michigan IV .... 42 — 161

30th Week
Carton Puzzles 43 80 162
No Splitting of Hairs! .... 43 — 164
Toy Soldiers—Halt! 44 80 165

31st Week
Dicey Question .... 44 — 166
Party Question 45 80 167
First and Last Day of the Year 45 81 169

32nd Week
Bridge Party 45 81 169
Beside/Below Redux 46 81 170
Weight Trick III 46 81 171

33rd Week
The Spy Who Came around the Corner 46 81 172
Grandfather and Grandson .... 47 — 173
Finally, Concerning Jealous Husbands .... 48 — 173

34th Week
One More Time: Knight on the Chess Board .... 48 — 174
Battle of Cards 48 81 175
Remains of a Long Division 49 82 177

35th Week
Ten Little Slips of Paper .... 49 — 179
Rook Walk 49 82 180
Married Couples 50 82 180

36th Week
Legal Problem from Antiquity 50 82 181
At the School Dance .... 50 — 181
Square Numbers 51 82 182

Black Belt

37th Week
Half-Truths .... 53 — 182
The Great Hunt .... 53 — 183

38th Week
Headscratcher with Caps 54 82 185
Schoolgirls .... 54 — 185

39th Week
Tennis Tournament 55 83 187
Hurdle Race 56 83 188

40th Week
Vacation Days 57 83 191
The Wine Grower's Estate .... 57 — 192

41st Week
Divine Injunction 58 83 193
Mathematicians in Conversation 59 84 195

42nd Week
Gifts 59 84 196
The Fickle Sultan 60 84 198

43rd Week
Military Band—Blow Your Own Horn! 60 84 198
At the Round Table 61 85 199

44th Week
WaterMonster .... 61 — 202
Tommy's Surprise 62 85 203

45th Week
A Number in Mind 63 86 204
The Hungry Sales Woman 63 86 206

46th Week
Another Fragmentary Division .... 64 — 207
Return of the Sultan 64 86 208

47th Week
An Interesting Game 65 86 209
At Lake Michigan V .... 65 — 209

48th Week
Computational Wizard .... 65 — 211
The Puzzling Scribble .... 66 — 211

49th Week
Footrace 66 87 213
No Tricks 67 87 213

50th Week
The Balance Scale .... 67 — 215
Unsuccessful Decree .... 68 — 215

51st Week
Book Lovers 68 87 216
A Party of Eight .... 69 — 217

52nd Week
Weighing Once Wins! 69 87 218
Clever Heirs .... 70 — 218

Preface

Many of us start our day with exercise. No wonder, we've always been told about the importance of regular fitness training—Rasmussen Reports found that over 80% of Americans believe that regular exercise is important. At the same time, very few of us concern ourselves with exercising our brain, the organ that makes us human. Many of us are even proud of this....

Telling riddles is one of my great passions. If I happen to hear an interesting brain teaser, I won't rest until I've passed it on to all my friends. Occasionally, though—to my great chagrin—I run into surprising resistance. If the puzzle is a bit longer, some people will break right into the middle of it and start talking about something else. "That reminds me that...." If the teaser is a short one, though, and makes a telling point, others will laugh and add mechanically, "Ah, a well-rested brain thought that one up!"

Yes, fortunately, there are still people with "well-rested brains."

Unfortunately, most people don't appreciate systematic thought and the power of new ideas. In school, it is math that's most responsible for developing our logical thinking. For the majority of students, however, math is a collection of formulas to be crammed for tests. Is it really necessary to demonstrate this? Just look at how the teaching of math in school is depicted in literature.

And beyond school? After graduation, "brain training," minimal to begin with, almost ceases, and we are fully occupied with the routines of everyday life.

We really don't know what all our intellects are capable of. Even absent a Rasmussen Report, we can carry out a small statistical survey among our own friends. Let's challenge them, one at a time, with the following simple brain teaser:

> "Someone called me on the phone," explains Jack, "and when I asked with whom I was speaking, the person calling was surprised I didn't recognize the voice, since the mother-in-law of his mother is my mother. I didn't believe what the caller was saying, since I have no siblings."
>
> What is the relationship between Jack and the caller?

Most people to whom I present this puzzle ask me to repeat it, to give them a chance to pull themselves together, or maybe change the subject. Some say confidently, "It was Jack's mother who called him, right?" And then I have to argue that this is wrong. It occurs only to a few of them to examine their answer and perhaps reach that high peak from which one doesn't just guess, but solves the problem through logical thinking.

And really, the answer is quite simple. (It helps to draw little circles on a paper to diagram the people and then lines between the circles to describe their relationships.) The mother-in-law of the caller's mother is Jack's mother; in other words, the husband of the caller's mother and Jack are children of the same mother. But since Jack has no siblings, the husband of the caller's mother is none other than Jack himself. That is, Jack is the husband of the caller's mother, and thus the caller is Jack's child (or perhaps, his wife's child).

Everybody is astonished: the caller, since the father hasn't recognized his child's voice, and the reader, seeing how simple the solution of such a puzzle can be.

"Really, even I could have solved that," you might say—and you would be correct. Try out the brain teasers in this book, and you'll see that you can solve them too.

What are the prerequisites for solving brain teasers?

First, you need to have good ones! Please browse through the book. I hope that even a picky reader will find ample material to his or her taste.

Second, you have to be committed to thinking logically. You have to think attentively and systematically through what you know, and then consider what conclusions can be drawn from it. This is not in and of itself a prescription for solving problems (no such prescription unfortunately—or thank God—exists), but rather it is a basic observation.

Third, you need an idea. Actually, not every brain teaser requires this. The problem just presented, for instance, unraveled itself, as it were, after a bit of logical reflection and without any brilliant ideas. With many brain teasers, however, you have to step a bit "outside the box," become mental "innovators." At first, of course, you require only modest ideas. But later, in proportion with the level of difficulty, the pleasure you feel at finding a good idea is correspondingly greater.

I have yet to find among my friends anyone who, when reading a book, is able to stick to the principles and instructions of the preface. Nevertheless, I would like to offer such advice. The reader can make use of it as he or she likes. Cheers!

As the reader will perhaps have noticed, this book is not organized into chapters, but into 52 weeks. I've organized the book this way so that the reader will find intellectual treats for every week of a year. The fact that each week presents only a few such treats (in the first 36 weeks, three per week; in the last 16 weeks, only two per week), counsels patience.

Take your time solving the problems. Don't run headlong to the solution, and don't pass the problem over lightly either, just because you think, "I'm not going to be able to solve this." Sure you will. You'll manage if you persist.

If you don't solve the problems in the order in which they are presented, it can well happen that you run into unfamiliar concepts. So I recommend that you solve the puzzles in order. Also, you should always look at the solution presented in the book, even when you solve the problem independently. Often, the solution we give introduces concepts you will need later on or presents general principles that can be profitably put to use in problems that come later. Also, if necessary, consult Section 1 of the Appendix, p. 221, surveying the terminology we use.

If after several attempts, you are still unsuccessful in solving a brain teaser (after repeatedly trying and reflecting on it for, let's say, three or four days), you should first look at the *hints* in the second part of the book. There you will find ideas for solutions to a number of problems. These are not complete solutions, which are given in the third part, *Solutions*. In *Hints*, you may also find references to earlier problems—where needed. The little "Hint-Man" greeting the reader in the margin indicates that there are ideas and concepts for the solution in *Hints*.

A separate title designates the last sixteen weeks: *Black Belt*. In this part, we pose only two brain teasers per week, but if you manage two such puzzles, you can give yourself a pat on the back. I recommend this section to all who get a kick out of experiencing beautiful ideas.

Off we go! Readers and solvers, he who seeks, finds. Look for the solution and find salvation! Have fun with it. Train Your Brain. Now it's time for our weekly brain teasers!

The Gym

1st Week

A Journey by Ship

Twice a day, at noon and midnight UT (Universal Time[1]), a ship leaves New York for Lisbon, and another leaves Lisbon for New York. The ships travel the same route, and passage takes exactly eight days.

Recently, I traveled on one of these ships from New York to Lisbon.

How many passing ships did I count? (I include in this count those arriving at my departure and those departing upon my arrival.)

Hint: p. 71 Solution: p. 89

Two Games of Chess

I've been an enthusiastic chess player for a long time, but I've never been able to get my sister to learn the game. She knows little more than the basic moves.

Two friends, Tom and Paul, came over for a visit, and, as was our custom, I played a game with each of them. I was in bad form, and I lost both games. When my sister came into the room and heard how I lost, she couldn't restrain herself:

"I'm ashamed of my brother's performance!"

[1] or Greenwich Mean Time, the mean solar time at the Royal Observatory in Greenwich, London.

"Please give me a chance" she said to my guests, "to even the score. I'd like to play a game with Paul, and another with you, Tom. I promise I'll get a better result than my brother. I won't even ask for the advantage of going first in both games. With Paul, I'll play white, and with Tom, black. I'll also give you a big advantage: I will play the two games simultaneously."

And that's how it was. Even though my guests put their best efforts into their games, my sister's record turned out to be better than mine.

How is this possible?

Hint: p. 71 Solution: p. 89

A Journey by Air

Boston is about 200 miles from New York City. We traveled in a small plane from Boston to New York City and back. The speed of the airplane was 150 mph.

We observed that during the entire flight, a strong, steady wind with a speed of 30 mph was blowing from Boston to New York City. So, from Boston to New York City, we were flying with the wind, which increased our speed. On the return trip, the wind reduced our speed correspondingly. Therefore, the wind did not affect how long it took to fly to New York and back to Boston.

Is our conclusion correct?

Hint: p. 71 Solution: p. 90

2nd Week

■■□□□

The Math Whiz

"It is difficult to subtract fractions in your head," said John.

"That's right," answered Peter, "but you know, there are several tricks that can help you. You often get fractions whose numerators are one less than their denominators, for instance,

$$\frac{1}{2}, \ \frac{3}{4}.$$

It's easy to figure out the difference between two such fractions. The trick is easy to understand:

$$\frac{3}{4} - \frac{2}{3} = \frac{4-3}{4 \times 3} = \frac{1}{12}.$$

Simple, right?"

Can you always use this method?

Solution: p. 91

The Will

A sheik is lying on his deathbed. He summons his two sons to tell them that he has hidden his treasure in a nearby oasis, and that in his will he stipulates that the sons are to travel to the oasis, and that the entire treasure shall belong to the son whose camel arrives second at the oasis.

Following the death of their father, the two sons are in a ticklish situation. Both wish to get the treasure. But how can a son arrange so that his camel will arrive later at the oasis than his brother's? If he would set up camp in the desert, then the other would do the same, and in the end, they would both die of thirst....

So they go to the Qadi to get his advice. The Qadi asks them to step close to him. The sons dismount their camels and obey the Qadi, who whispers something in their ears. Whereupon they both mount the camels

and ride furiously in the direction of the oasis. And in the end, this breakneck dash wins one of them the treasure.

What advice did the Qadi give them?

Solution: p. 92

The Clock

In the last century, at a time when clocks were mechanical (most of them clocks with springs), a man told the following story:

"I don't have a wristwatch, but in my apartment, there is a marvelous and accurate wall clock. Sometimes, though, I forget to wind the clock, and I don't have a radio to help set it correctly. Once, as I was leaving to visit a friend, I noticed that my wall clock had stopped. I had time to wind the clock and get it going again. I spent the evening at my friend's. On the radio, we listened to a wonderful concert, and then I went home and set my clock accurately."

How could he set the clock with the correct time, even though he never measured the time it took to walk from his friend's apartment to his own?

Hint: p. 71 Solution: p. 92

3rd Week

Nicholas's House

"Daddy, in school today we were given the assignment to draw this house with one continuous line without lifting the pencil from the paper, and never drawing over a line twice."

"And did you manage it?"

"Yes, after a few tries, I got it."

"Well then, let me ask you a couple of questions. From which point do you have to start to get a solution? At which point do you end? And if the starting point is correct, what rules do you have to follow to draw the house? If you answer these questions correctly, you'll be able to draw the house every time without any trial and error."

Can you answer these easy questions?

Solution: p. 92

An Unusual Year

"I was born in an unusual year," tells Kate, "Fridays fell on the 13th three times. Not only that—it was a leap year."

"When is your birthday?"

"I was born on the first of April."

"And by any chance was the first of April a Friday that year?"

Answer for Kate.

Hint: p. 72 Solution: p. 93

Competitors, or *The Raise*

There were three young applicants for a job. To decide which one to hire, the boss asked them:

"The starting monthly salary is two thousand dollars. If your work is good, you get regular raises. You have a choice: We can pay you monthly and raise your paychecks by $150 every month; or we can pay you twice a month, and raise each of your paychecks by $50. Which do you prefer?"

Two of the applicants immediately chose the first option, while the third, after thinking about it for a short while, decided on the second option. The boss hired the third applicant.

Why?

Hint: p. 72 Solution: p. 94

4th Week

■■■■□□□

The Metal Tube

A brass tube weighs 85 lb. I want to cut it into two pieces, so that the weight of each piece is an (integer[2]) multiple of 1 lb.

---

[2]For the basic terminology, see Section 1 of the Appendix, p. 221.

In how many ways can I do this?

Solution: p. 95

The Ancient Greek Copper Lion

An ancient Greek copper lion tells the tale:

"I am a fountain; there is a basin in front of me. There are orifices hidden at four places in my body: in both of my eyes, in my mouth, and in the hollow of my right knee. My right eye can fill the basin in two days; my left eye can do it in three days; my right knee in four days, whereas my mouth needs just six hours to do it. (Recall that for the ancient Greeks, a day was divided into 12 hours!)

"How long does it take to fill the basin if water flows from all four of my orifices at the same time?"

Hint: p. 72

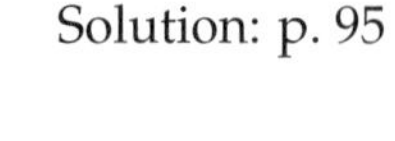

Solution: p. 95

The Arab

Before his death, the venerable Arab arranged to leave to his three sons his herd of 19 thoroughbred horses. The eldest son was to inherit half of the horses, the middle son one quarter, and the youngest one fifth.

After their father's death, the three sons did not know how to carry out his will. No wonder; they wouldn't want to butcher thoroughbreds! So they asked the advice of the wise Qadi.

The Qadi offered a solution that pleased the sons and did not butcher any of the thoroughbreds.

What did he suggest?

Hint: p. 72 Solution: p. 96

5th Week

■■■■■□□□

On the Streetcar

Midway through the last century, in the 1950s, there were two types of tickets for adults in the Budapest streetcar system: a transfer ticket for 70 fillérs and a line ticket for 50 fillérs. You had to buy the tickets from the conductor, and there were 5-fillér coins, 10-fillér coins, 20-fillér coins, 50-fillér coins, and 1-forint coins (100 fillérs make a forint).[3]

One afternoon, an adult boarded a streetcar and handed the conductor one forint. Without saying a word, the conductor, who had never seen the passenger before, handed him 30 fillérs and a transfer ticket.

How did the conductor know the passenger wanted a transfer ticket?

Hint: p. 72 Solution: p. 96

Hunting Hares

Recently, I participated in a hare hunt. I observed as one of the beaters flushed a hare from a bush. The hunters and the greyhounds did not immediately notice, so the hare had a head start of 34 hare-jumps before one of the greyhounds started chasing it.

Of course, a greyhound takes much longer jumps than a hare. To be precise, five greyhound-jumps equal nine hare-jumps. On the other hand, a hare jumps more often than a greyhound: for eight greyhound-jumps, the hare can make eleven.

The greyhound caught the hare with ease. How many jumps did the hare make before the greyhound caught him?

Hint: p. 73 Solution: p. 96

Telephone Cable—A Long Line?

Let's assume the earth is a perfect sphere, and the equator measures 25,000 miles.

A telephone company installed a telephone line at a height of 10 feet above ground around the entire equator. Considering the frequency of damage to the line, a company engineer proposes raising the entire line by 2 feet. A company executive disagrees; the additional cable required

---

[3]The Hungarian unit of currency is the *forint* (named after the Italian town of Florence, where the coins were first minted many centuries ago). When this book was originally written, coins of those designations still existed. The fillér coins were recalled in 1999. The 1-forint coins and the 2-forint coins were withdrawn from circulation in 2008. Today, about 200 forints convert to a dollar.

would prove very expensive, since 1 foot of cable costs $20. The engineer offers to pay for the extra cable out of his own pocket.

How much would this offer cost the engineer?

Hint: p. 73 Solution: p. 97

6th Week

Dinner of Dumplings

Three exhausted tourists arrive at an inn. They sit down at a table and order a large plate of potsticker dumplings. By the time the host brings the food, all three have fallen fast asleep. Later, one of the tourists wakes up, eats one third of the dumplings, and falls asleep again. Then the second wakes up; not noticing that one of his companions has already eaten, he eats one third of the dumplings that are left, and then also goes back to sleep. Finally, the third tourist wakes up, and eats one third of the dumplings on the plate. When the host returns the next morning, he finds eight dumplings on the plate.

How many dumplings did the host serve?

Solution: p. 97

The Paradox of Protagoras

Protagoras, one of the greatest of the Greek philosophers, lived and worked in the fifth century BCE. At one time, he taught jurisprudence. He reached the following agreement with one of his students: the student was required to pay his tuition, or not to pay it, according to whether he won his first trial.

The student finished his course of study, but chose not to begin the practice of law. Protagoras sued his student for refusing to pay his tuition fee.

Was this the right course of action?

Solution: p. 98

Say How Many Flags Are Flying

In class, the students are designing flags. Each flag is divided into three parts by two vertical lines (just like the Italian flag[4]). The children color the three stripes with red, blue, green, or yellow.

(a) How many different flags can they design if the colors in a flag may not repeat?

(b) How many flags are possible if the two outside stripes can be of the same color?

Solution: p. 99

7th Week

■■■■■■■□□□

Homework

My son's homework assignment is two multiplication problems. In the first, a six-digit number ending in 7 has to be multiplied by 5; to his surprise, he finds that the result could be obtained from the original number just by moving the 7 from the end of the number to the start.

His surprise is even greater when it turns out that the second problem is similar. In the second multiplication, a six-digit number beginning with 1 has to be multiplied by 3; the result could be found by moving the 1 from the beginning to the end.

What was his homework?

Solution: p. 102

The Superhotel

Seven tired men enter a country inn. They request lodging, but stipulate that each must have a single room.

[4]The Italian flag consists of three vertical stripes of the colors green (left), white (middle), and red (right).

The innkeeper informs them that he has only six rooms vacant, but hopes nonetheless to lodge his esteemed guests in accordance to their wishes.

He leads the first guest into the first room, and asks another to wait a few minutes in the first room. Meanwhile, he brings the third guest into the second room, the fourth guest into the third room, the fifth guest into the fourth room, and the sixth guest into the fifth room. Then, he returns to the seventh guest in the first room and leads him into the sixth room. In this manner, he accommodated them all comfortably.

Or maybe not?

Solution: p. 102

Jealous Husbands

Two jealous husbands wish to cross the river with their wives, but there is only one two-person boat available to them. Neither one of the two men wants to leave his wife alone with the other man.

How can they manage the crossing?

Solution: p. 102

8th Week

■■■■■■■■□□

Another Superhotel

Three tourists sought lodging in a small American hotel. The proprietor offered them a three-room suite for $300. The three men went upstairs with the bellhop to have a look at the suite. Everything seemed fine, and each tourist gave $100 to the bellhop. When the proprietor received the

$300 from the bellhop, he discovered that he had made a mistake: the price of the suite was only $250. Therefore, he sent the bellhop back with five $10 bills. On his way upstairs, the bellhop realized it would be a problem to distribute the five $10 bills among three people. So he slipped two of the $10 bills into his own pocket, and gave back $10 to each of the three men. Thus, they had each paid $90, which adds up to $270. Twenty dollars ended up in the bellhop's pocket. Altogether, that makes $270 + $20 = $290, even though the three guests had originally paid $300.

What happened to the missing ten dollars?

Hint: p. 73 Solution: p. 103

Three-Letter Words

In these two consecutive sums, we add two three-digit numbers and then add a three-digit number to the result. The digits are represented by letters. Repeated letters represent the same digit; different letters represent different digits.

What digits do the letters represent?

Hint: p. 73 Solution: p. 104

In the Kindergarten

By coincidence, all the children in one kindergarten have only two hair colors (brown and black) and two eye colors (blue and green). (We assume that not all the children have the same hair color or the same eye color.)

For a game, the teacher wants to choose two children with different hair colors and eye colors.

Is it certain that she can find two such children?

Hint: p. 73 Solution: p. 105

9th Week

■■■■■■■■■□□□

How Old Is Susie?

"It's interesting," says Susie, "that my mother is exactly half as old as my father and I together. Between them, my father and my mother are 100 years old, and the age of each is a prime number."

How old is Susie?

Hint: p. 73 Solution: p. 106

Curious Addition

"My Dear," says a woman to her husband, a famous mathematician, "I found this piece of paper under your desk. Do you still need it?"

"Oh, yes, I do, indeed," the Professor replies, "give it to me, please!"

"Is this supposed to be an addition?"

"Yes, of course."

"In that case, you'd better recalculate it. You added it up wrong."

"No, absolutely not. This is an addition . . . ."

Why does the Professor think that his addition is correct?

Hint: p. 74 Solution: p. 106

Exercise in Logic

Let us consider the following statements:

(1) Bacon wrote all the plays of Shakespeare.

(2) There is a play of Shakespeare's that Bacon wrote, but Bacon did not write all of Shakespeare's plays.

(3) There is a play of Shakespeare's that was not written by Bacon.

We do not know which of these statements are true and which are not—and we may never know with certainty. Of these three statements, however, there are two that can both be simultaneously true and both be simultaneously false. Among the three there are also two statements that can both be simultaneously false but not simultaneously true.

Which are these statements?

Solution: p. 111

10th Week

The Local Lurches On

From Richmond, the local train travels at a speed of 30 mph in the direction of Washington, DC. Leaving somewhat later, also from Richmond, the express train travels at a speed of 60 mph. Normally, the express train catches up with the local at Washington, DC. But due to a mechanical failure, the local train, after traveling 2/3 of the distance, is able to continue at only half of its original speed. For this reason, the express has already caught up with it $27\frac{1}{9}$ miles before Washington, DC.

How many miles is it from Richmond to Washington, DC?

Hint: p. 74 Solution: p. 111

Cocoa Island

Two nations, Bodo and Wizzer, live on Cocoa Island. They have lived on the island so long that a newcomer cannot distinguish them. The two nations are distinct from one another in one important trait: a Bodo always tells the truth, while a Wizzer always lies.

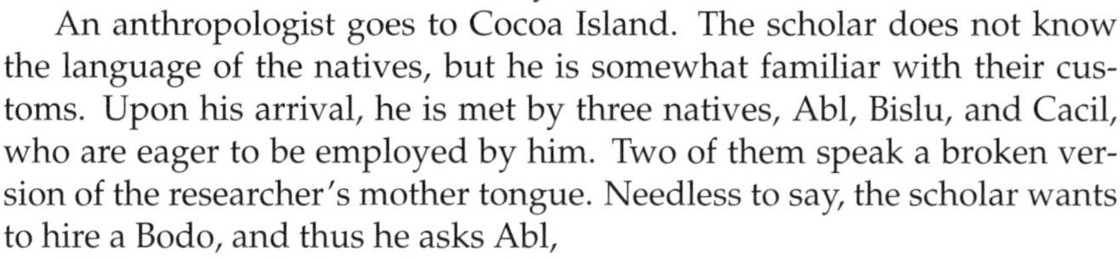

An anthropologist goes to Cocoa Island. The scholar does not know the language of the natives, but he is somewhat familiar with their customs. Upon his arrival, he is met by three natives, Abl, Bislu, and Cacil, who are eager to be employed by him. Two of them speak a broken version of the researcher's mother tongue. Needless to say, the scholar wants to hire a Bodo, and thus he asks Abl,

"Abl, are you Bodo or Wizzer?"

"Bhio, fa kutja marjion," is the reply.

"Well, I didn't understand a word of that. Bislu, can you tell me what Abl said?"

"Abl say he is Wizzer," answers Bislu.

"And what do you say he said, Cacil?"

"Abl say he is Bodo," answers Cacil.

Can we determine whether Bislu and Cacil are Bodo or Wizzer?

Hint: p. 74 Solution: p. 112

The Little Troublemaker

Alex tells us, "I was seated at my chess table; my son and daughter sat next to me. My daughter was working on an arithmetic assignment—the exercise was to carry out a number of divisions. When she left the room for a moment, her little brother busied himself with covering up the digits in the divisions with chess pieces.

"When I glanced over, only two digits were still visible. Here's what I saw:

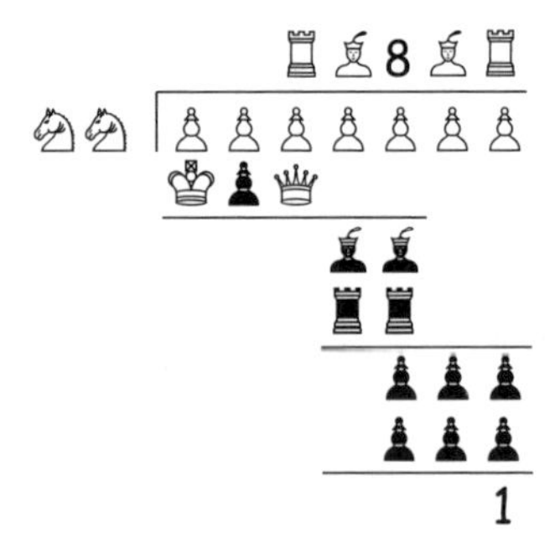

"Had I just swept away the chess pieces, my son would have howled in protest; had I left them in place, my daughter would have pouted over the disruption of her homework. I had no choice but to figure out the division without moving the chess pieces. When my daughter returned, I wrote down for her on a separate piece of paper her progress with the division."

As you see, Alex was a good puzzle solver—and he avoided a family quarrel.

In his position, could we have done this too?

Hint: p. 74 Solution: p. 112

11th Week

In the Garden

Mr. Klein entertains guests. After a pleasant afternoon coffee, he leads them outside into the marvelous garden of his villa. The guests ask that he show them the beautiful garden in its entirety. Mr. Klein, anxious not to lose too much time, wishes to lead his guests around so that they travel each path exactly once.

Can he do this? And if he can, down which staircase should they enter the garden?

Solution: p. 113

Five Little Decks of Cards

A magician and four enthusiastic fans sit at a table. The magician speaks:

"I'd like to demonstrate a simple trick for you, and I hope you will be able to repeat it in short order."

He takes out five small decks of cards, each deck consisting of five cards. Now he instructs each member of his audience to choose a deck. They pick up the decks they chose. Each person then memorizes one of the cards in the deck he picked. The magician collects the five decks, divides them up into five decks of five cards, but in a different way, and says "I'll now show you the five decks in order, spreading each out like a fan. As soon as you see the card you memorized, let me know, and I'll tell you which card you chose."

And so it transpired. Even when one of the spread out decks held the chosen cards of two of his fans, he accurately identified which card was picked by whom.

Can we repeat this trick?

Hint: p. 74 Solution: p. 113

A Clumsy Division

"That recent division problem (10th Week, "The Little Troublemaker," p. 14) was great; it pleased me so much," says Karli, "that I immediately started making one of my own. I finally succeeded:

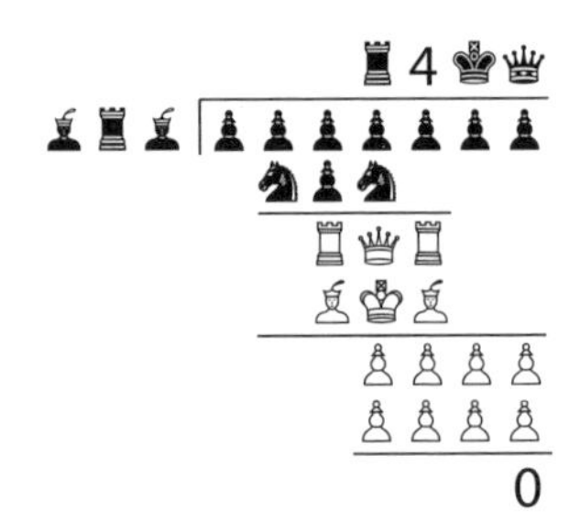

Unfortunately, this has several solutions, so I'll need to tell you a few more things.

"So, I'll tell you this: If you divide the divisor by 9, it leaves the remainder 7. On the other hand, if you divide the quotient by 9, it leaves the remainder 3. And be aware that identical chess pieces don't necessarily represent the same digits."

Given all this information, can we solve the puzzle Karli dreamed up?

Hint: p. 75 Solution: p. 113

12th Week

Cycle Tour I

Steve and Tommy, two good friends, went on a cycling tour. Tired of traveling long stretches of road in silence, they decided to pass the time with games that can be played while cycling. Tommy proposed the following:

"Steve, I'm sure you know this game. The starting player says a number between one and ten (that is, a number not less than one and not greater than ten). Then we take turns adding a number between one and ten to the number our opponent has just named, saying the resulting sum as our answer. The winner is whoever reaches 100 first."

"Sure, I know that one. Somebody once explained to me the trick that helps you win," Steve answered; "I just hope I can manage to remember it."

And as it turned out, it wasn't long before he remembered the trick.

What is the trick?

Hint: p. 75 Solution: p. 114

More Jealous Husbands

Three jealous husbands want to cross a river with their wives in a two-person boat. None of the men, though, dares leave his wife alone with another man.

Arrange the river crossing.

Solution: p. 114

Toy Soldiers—On the Double!

A little boy is playing with his toy soldiers. He carefully sets up his soldiers in rows so that each row has the same number of soldiers. But there are always some soldiers left over.

He complains to his father. If he sets up his soldiers in four rows, three soldiers are left over; if he sets them up in nine rows, five soldiers are left over. Then he asks his father if a lot of soldiers will be left over if he tries setting them up in 36 rows.

His father couldn't answer the question right off the bat.

Let's help him!

Hint: p. 75 Solution: p. 115

13th Week

Big Problem—Chess Problem

I tried to figure out how many ways I could set up a white and a black rook so that neither could take the other.

I tried many options and saw it would be a long time before I solved this problem.

The reader may prove to be more clever than I am in calculating the possibilities.

Hint: p. 75 Solution: p. 116

Cycle Tour II

Steve racked his brain over how he might outwit Tommy. They had figured out (12th Week, "Cycle Tour I," p. 16) that the starting player always wins if he follows the right strategy.

Steve wanted to change the rules so that the starting player should lose. Steve proposed lowering the number 10, the upper limit of the starting number and of what can be added (to the number reached by the opponent), so that the starting player would always lose if his opponent followed the best strategy.

To keep the game from dragging on too long, however, he wanted to set this limit to five or higher.

What number should we suggest to Steve?

Hint: p. 75 Solution: p. 116

With Time and Patience, You'll Go Far

Express the number 100 in as many ways as possible using only the digits

$$1,\ 2,\ 3,\ 4,\ 5,\ 6,\ 7,\ 8,\ 9$$

and only the four arithmetical operations (addition, subtraction, multiplication, and division). Each of the digits must be used exactly once.

Two examples are

$$100 = 97 + \frac{6}{4} + \frac{1}{2} + \frac{3+5}{8},$$

$$100 = 123 + 4 - 5 + 67 - 89.$$

Who has the most patience and can produce the most expressions like this?

In addition to these two examples, we can come up with another 31. And there are surely more.

Solution: p. 116

14th Week

Pouring Judiciously

We have an 8-quart container, a 5-quart container, and a 3-quart container. The 8-quart container is completely filled with wine.

How can we divide the wine into two equal parts if we cannot use anything other than these three containers?

Solution: p. 119

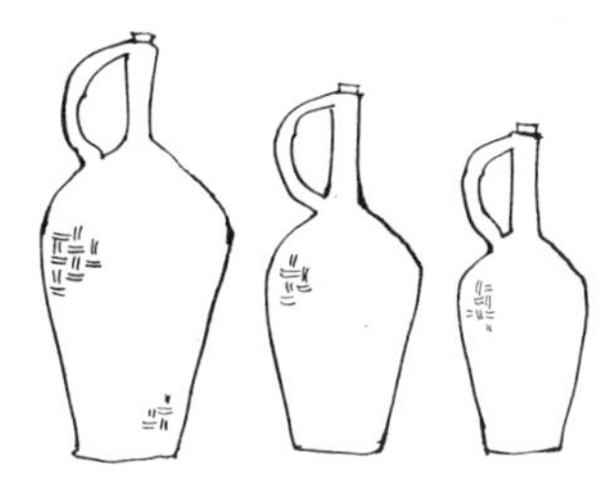

The Barber

A stranger came into town and started a conversation with the hotel barber.

"Do you have a lot of competition?" he asked.

The barber replied:

"Oh no. Not at all. There is no other barber in town. I shave all the men who don't shave themselves. But so that I don't have more work than I can handle, I don't ever shave anybody who shaves himself."

The stranger looked at the barber, laughed, and said, "There's a contradiction in what you say."

The barber pondered this; try as he might, he couldn't see any contradiction in his words.

Can we help him?

Hint: p. 75 Solution: p. 120

Council of Elders

The nine men of the Council of Elders are named:

> Potter, Baker, Smith, Carpenter, Cook, Tailor, Carter, Mason, and Gardener.

Their trades are

> *potter, baker, smith, carpenter, cook, tailor, carter, mason,* and *gardener.*

Of course, these men may not practice the trade whose name they happen to carry.

Here are some news items and rumors from the village:

The *cook* is the father-in-law of the *baker*. Mr. Cook is engaged to the only daughter of the *carpenter*, who has already turned down two of Cook's rivals, the *smith* and the *potter*, who remain unmarried. Tailor's daughter is the outstanding tennis partner of her fiancé.

For the leadership of the local chess club, Potter—who is a bachelor—has enlisted the only member of the Council of Elders who practices the trade of his name. Carter and his son-in-law play Lotto together. The *potter* has only one sister and no other siblings; his sister is the wife of Smith. The *gardener* and the *tailor* have established their joint business along family lines: they have each married the other's sister. Two members of the Council of Elders have daughters, no member has a son, and no member has more than one daughter. The name of the *gardener* is the same as the trade of the member whose name is the same as Tailor's trade. Moreover, the name of the *tailor* is the same as the trade of the member whose name is the same as Carter's trade.

It must now be crystal clear which trade each man practices! Or maybe not? Let's think it through!

Solution: p. 121

15th Week

The Story of Josephus Flavius

In the year 66 BCE, the Romans conquered the city of Jotapata. The Jews were forced to flee after defending the city for 47 days. Among their number was the great chronicler of history, Josephus Flavius, who, together with his 40 companions, took refuge in a cave. They all agreed that they would rather die than fall into the hands of the Romans. Josephus Flavius and one of his friends secretly disagreed with this decision.

The following proposal of the historian was adopted. Josephus Flavius arranged the 41 people in a large circle; each third person would be killed until only one remained, who would then commit suicide. Josephus Flavius placed himself and his friend in the circle so that they would be the last two left alive. Thus, they escaped.

Where did Josephus Flavius and his friend stand in the circle?

Solution: p. 122

Car Trip

We took a long trip by car. A separate truck transported our luggage.

Since the truck travels more slowly, we sent it on its way earlier, and agreed to meet in Chicago. The car traveled at a speed of 66 mph, and we arrived in Chicago at 6:53. The truck traveled at 42 mph and, notwithstanding its head start, arrived only at 7:11.

How many miles before Chicago did we pass the truck?

Hint: p. 76 Solution: p. 122

The Importance of Plain Speaking

I asked a friend how old the members of his family were. He answered me as follows:

"In six years, my father will be three times as old as I was when the number of years he had been alive equaled the sum of my age and the age of my younger sister at the time. I am now exactly the age my father was then. In nineteen years, my father will be twice as old as my younger sister is today."

Do we now know how old my friend's family members are?

Hint: p. 76 Solution: p. 123

16th Week

The Fly

Appleton is 120 miles from Brownsville. There is a straight highway with a bike path connecting them. A racing team of cyclists from Appleton starts off towards Brownsville. At the same time, several amateur cyclists from Brownsville leave for Appleton. The racing team travels at 25 mph and the amateurs at 15 mph.

At the moment the racing cyclists from Appleton start, a fly takes off from the starting line, also heading towards Brownsville, flying at a speed of 30 mph, until it reaches the group of cyclists from Brownsville. At this

point, it reverses course, and flies back towards the Appleton group. In this manner, it flies back and forth between the two groups, and tragically perishes when the two teams accidentally collide with each other.

How many miles has our heroic fly flown?

Hint: p. 76 Solution: p. 125

Astonishing Trick

I take out a clock and instruct my partner to think of a number between 1 and 12. Next, I grab a pencil and use it to tap the numbers on the face of the clock, here and there. While I am doing this, I ask my partner to add 1 to the number he chose each time I tap a number on the dial, until he reaches 20, when he should shout "twenty." He will be really surprised to see that when he shouts "twenty," I will have just tapped on the number he chose.

How can I do this?

Hint: p. 76 Solution: p. 125

Prize Question

Thirty students participated in a math contest. They had to solve three problems. I can recall the following about the results:

- 20 students solved the first problem.
- 16 students solved the second problem.
- 10 students solved the third problem.
- 11 students solved the first and second problems.
- 7 students solved the first and third problems.
- 5 students solved the second and third problems.
- 4 students solved all three problems.

I want to know how many participants did not solve a single problem.

Can you help me?

Solution: p. 126

17th Week

Bicycle Motor

A cycling club keeps regular daily records of the combined total miles its members ride. The club was commissioned to test a motor that can be mounted onto a bicycle. First, the motor was mounted on 40 bicycles; this increased the daily mileage by 20%. Then, the motor was mounted on 60% of the club's bicycles (including the 40 bicycles on which the motor was first mounted); as a result, the daily mileage increased to 2.5 times the original daily mileage.

How many cyclists are in the club? By what factor would the daily mileage increase if the motor is mounted on all the bicycles?

Solution: p. 128

Each Ninth Drops Out

A group of 35 people decides to play a new party game; the game, however, requires only 18 players. So they arrange themselves in a large circle and begin counting from one. Whoever counts a 9 or a multiple of 9 is out. They continue this until only 18 people are left.

One of the men arranged the group in a circle so cleverly that just he and his 17 close friends remained in the game.

How did he do this?

Solution: p. 129

Impressive Card Trick

We select 27 cards from the deck and ask a volunteer from the audience to memorize one of them. Then we shuffle the cards and lay them out, face up, in three decks on the table as follows: the first card goes in the first deck, the second card in the second deck, the third card in the third deck, the fourth card back in the first deck, next to the first card, and so on.

Once we have laid out the cards, we ask our volunteer, who memorized the card, to tell us in which deck the card is.

Then we put the three decks together so that the deck with the memorized card is in the middle. We repeat this twice more—that is, we lay the cards out on the table as described above, ask the volunteer to tell us in which deck the card is, and then gather the cards by putting the deck with the memorized card between the other two.

Then we take all 27 cards in hand. The chosen card is—counted from the top—the 14th card.

Can we explain why this is so?

Hint: p. 76 Solution: p. 129

18th Week

How Did I Know That?

"Think of a three-digit number such that the first and the last digits differ by at least two."

"I've done it." (She chose 246.)

"Reverse the order of the digits, and subtract the smaller of the two numbers from the larger."

"OK. I've done that too." (She got $642 - 246 = 396$ as a result.)

"Reverse the order of the digits of that number and add those two together."

"That's also done." (She got $396 + 693 = 1{,}089$.)

"The result is $1{,}089$. Right?"

"Yes. How did you know that?"

Hint: p. 77 Solution: p. 130

Weight Trick I

I have four boxes on my right, three of which weigh exactly the same. The fourth box has a different weight, however. I also have three boxes on my left, each weighing the same as the three boxes of equal weight on my right.

Using a balance scale only twice, I need to determine which one of the four boxes on my right has a different weight from the others, and whether

this box is heavier or lighter than the others. When I do the weighing, I may not use any weights, and I have only one balance scale at my disposal.

Solution: p. 131

Knight on the Chess Board

We would like to move a knight across the chess board so that we go from the lower-left corner to the upper-right corner, landing in the process on each square exactly once.

Is this problem solvable?

Hint: p. 77 Solution: p. 131

19th Week

Pocket Change

We ask a friend to take coins in each hand, so that in one hand he has an even number of coins and in the other an odd number. Now we instruct him to multiply the number of coins in his left hand by any even number, and to multiply the number of coins in his right hand by any odd number. Finally, we ask him to add the two numbers he obtained. When he tells us the sum, we can say at once in which hand he held an even number of coins and in which an odd number of coins.

How is this possible?

Solution: p. 132

Water Pipes

Mr. Smith has a nice, large yard. His property is laid out as a triangle, and the three sides are of equal length. He is quite happy with his land except that there is no well on it. On his own, Mr. Smith cannot afford to have a well dug.

So he agrees with his three neighbors (on the other side of the three dividing walls) that a well be dug, equipped with an electrical pump, and the water will be routed to his neighbors through pipes to the dividing walls of the yard. In this way, all four neighbors will have access to water.

They have already worked out the details, but Mr. Smith is still racking his brain over where to dig the well so that the water can be routed to his neighbors by using the minimal length of pipe. Mr. Smith's wife suggests digging the well in a corner of the property, since then they will need only a single length of pipe. One of the neighbors, on the other hand, thinks a well in the middle of the yard is the most economical, because the well would be of the same short distance from the three walls.

What counsel can we offer Mr. Smith?

Solution: p. 132

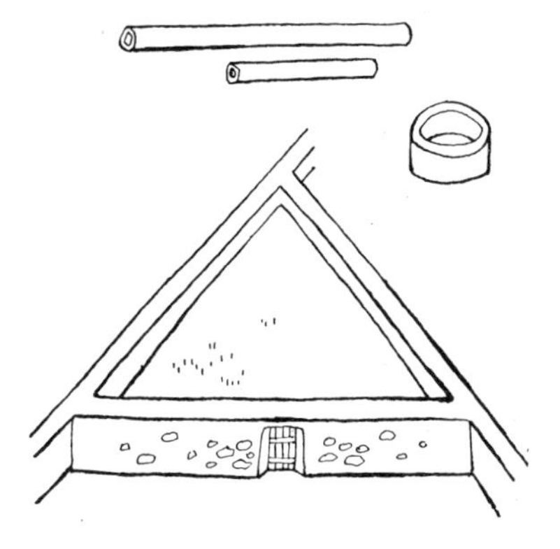

Cycle Tour III

"I have a great idea," Tommy said. "Let's change the rules of the game so that instead of winning by reaching 100 first, whoever gets to 100 first loses."

"That's fine," said Steve. "Just remind me of the ground rules; I don't quite remember them."

"You or I start by naming a number between 1 and 10 (1 and 10 included). Then by turns, we add a number between 1 and 10 (1 and 10 included) to the number the other has named and announce the result as our answer. The loser is whoever is forced to reach 100 (to be precise, at least 100) first. I'll start now," said Tommy, thinking how advantageous it would be to go first.

Was Tommy correct?

Solution: p. 133

20th Week

The Cans—Yes, We Can!

I have a 7-quart can and a 13-quart can; both are full of wine. I want to pour out 10 quarts of wine for my friend, but I have only a 19-quart empty can.

Can I measure out 10 quarts of wine with only these three cans?

Solution: p. 134

Two Puzzles in Two Languages

In the following three additions, digits should replace the letters so that the same letter is always replaced by the same digit, while different letters are replaced by different digits.

The following English puzzle is based on a cry for help:

```
   S E N D
 + M O R E
 ---------
 M O N E Y
```

One more puzzle of this type, a Hungarian one, comes from the rallying cry "LONG LIVE THE FIRST OF MAY" of the International Day of Labor (in Hungarian, "ÉLJEN MÁJUS ELSEJE!"):

```
   É L J E N
 + M Á J U S
 -----------
 E L S E J E
```

Since the second problem would have several solutions, we further assume that the sums of the digits of each of the two summands (that is, in ÉLJEN and in MÁJUS) are equal.

Hint: p. 77 Solution: p. 135

Shopping for Chocolate

Peter and Paul, along with their girlfriends, Theresa and Betty, bought some chocolate bars.

When they finished shopping, Peter noticed that for each chocolate bar, he paid as many dollars as the number of chocolate bars he purchased. The friends were really surprised that the same was true for Paul, Theresa, and Betty.

Each couple spent $65. Peter bought one bar more than Theresa; Betty, on the other hand, bought only one bar.

What is the name of Betty's boyfriend?

Hint: p. 77 Solution: p. 137

21st Week

Instant Addition

We get tremendous results with the following trick.

We ask a friend to think of two numbers—to keep it quick, two numbers under ten. He should write the numbers one under the other, add them up, and write down the result as the third number. Then he should add the second and the third numbers, yielding the fourth number. He should continue until he has ten numbers.

Then he should fold the paper between the fifth and the sixth numbers, and show me only the lower half of the paper (the half, in other words, on which the sixth through tenth numbers are written). After only brief reflection, I tell him the sum of his ten numbers and ask him to write down this sum. Then I watch, smiling, as he laboriously adds up the ten numbers to check if I am correct.

For instance, suppose he picked 8 and 5. Then he obtains the following numbers:

$$8, 5, 13, 18, 31, 49, 80, 129, 209, 338$$

The lower half of the paper reveals to me the numbers

$$49, 80, 129, 209, 338$$

I tell him at once that the sum is 880.

Can you repeat this trick?

Hint: p. 77 Solution: p. 138

Still More Jealous Husbands

We already know the story of the jealous husbands and the two-person boat (7th Week, "Jealous Husbands," p. 10, and 12th Week, "More Jealous Husbands," p. 17).

On this excursion, more than three couples are taking part. They are aware that previously two and then three couples have succeeded in crossing the river. They also hope to succeed.

Is their hope justified?

Hint: p. 77 Solution: p. 140

Soccer

Four soccer teams—the New England Revolution, the LA Galaxy, DC United, and the Kansas City Wizards—participated in a simple round-robin tournament (each team played every other team once). As is customary, the winner of a match received two points, and in the case of a tie, each team received one point.

The Revolution won the tournament with five points, DC United received three points, and the Galaxy received one point. Altogether, there were 11 goals; DC United scored five of them. DC United defeated the Galaxy 2 : 1.

What was the outcome of the match between the Revolution and the Galaxy?

Hint: p. 78 Solution: p. 141

22nd Week

■■■■■■■■■■■■■■■■■■■■■■□□□□□□□□□□□□□□□□□□□□□□□□□□□□□□

Prime Number

"Think of a prime number greater than three."

"Done." (11)

"Square it and add 17."

"OK." (138)

"Divide it by 12, and take note of the remainder."

"Yes, I've done that." ($138 = 11 \cdot 12 + 6$, so, the remainder is 6.)

"The remainder is 6. Right?"

"Yes! How did you know that?"

Hint: p. 78 Solution: p. 142

Cycle Tour IV

"I'm tired of just playing variations of the same game all the time. By now I know all the tricks," Steve complained.

"Agreed," Tommy replies. "We'll be setting up camp at Lake Michigan soon, and I have another game to share then. Meanwhile, though, I have one more variant for the game we've just about worn out."

Tommy suggests revising the rules of the game again:

1. Tommy picks a number between 1 and 10. This will be the starting number, and when it's his turn, he will add a number to *Steve's number* between 1 and the number he picked.

2. Steve picks a number between 1 and p. (We will determine the number p later.)

3. They play the game as before:

   (a) Tommy starts with the number he picked between 1 and 10;
   (b) Steve adds a number between 1 and p;
   (c) Tommy adds a number between 1 and the number he picked;
   (d) and so on;
   (e) whoever reaches 100 first wins.

Tommy says, "We'll take turns starting.

"My question is: how large should p be for the game to be fair?"

Will this game offer anything new?

Hint: p. 78 Solution: p. 143

Green Cross

The director seeks an assistant. From among the many applicants for the job, he selects three intelligent young men. He calls them into his office, and tells them:

"All three of you are intelligent young men. I'm going to pick the most suitable candidate by using an interesting logic problem: I have two small pieces of chalk in my hand, one green and one white. I'll turn off the light, and draw a small cross on each of your foreheads using the green or the white chalk. When the light is turned on, look at the foreheads of the other two candidates; if you see a green cross, raise your hand. As soon as you know the color of the cross on your forehead, say 'I know.' If you give a convincing explanation, you get the job."

After checking that they understood the instructions, the director turned out the lights and drew a cross with the green chalk on the foreheads of all three candidates. When the light was turned on again, all three

raised their hands, and after a brief pause, one of them, Mr. Sharp Logic, said, "I know."

"Great," answered the director, "and what is the color of the cross on your forehead?"

"It is green, sir."

How did Mr. Sharp Logic know the color of the cross?

Hint: p. 78 Solution: p. 144

23rd Week

Money for Ice Cream

This story took place a long time ago, when a scoop of ice cream cost 10 cents.

"I told my cousins about our uncle, a very nice but quirky person; we, the children, liked him a lot because he always gave us money for ice cream.

"One day, the four of us were visiting him, and he distributed 19 dimes among us. Actually, the way he did it wasn't quite fair, but we all got enough for a scoop. Our uncle wrote a two-digit number on a piece of paper, and then multiplied this number by itself, so he got a four-digit

number. The smallest of us got as many dimes as the first digit, the next smallest as the second digit, and so on (reading the digits from left to right). The two biggest children got the same amount. Being my uncle's favorite child, I got the most.

"How many scoops of ice cream was I able to buy with the money I got?"

Hint: p. 78 Solution: p. 145

At Lake Michigan I

Steve and Tommy arrived at Lake Michigan and set up camp. Now they could pass the time playing games that didn't need to be played in their heads.

Steve took out a box of matches.

"I once heard of the following game," said Steve, as he laid out the matches into piles of varying sizes. He told Tommy that the rules are as follows:

- The players alternate picking up the matches.
- In one turn, a player must pick up matches from *one* pile; he has to pick up at least one match or any part of the pile; if he chooses to, he can take the whole pile.
- Whoever picks up the last match loses.

They tried the game at once. To start the game, Steve arranged two piles of three matches each. Steve took two matches from one pile; Tommy picked up the remaining match. Then, beaming, Steve picked up two matches from the other pile, and Tommy saw that he was forced to pick up the last match. So Tommy lost.

Then they played a new game. Once again, Steve arranged the matches: each pile was a single match, except for one pile, which contained several matches.

Tommy goes first. What must he do to win?

Solution: p. 146

The Envious Cousin

It seems my cousin envied me for the big pile of dimes I got for ice cream (this week, "Money for Ice Cream," p. 32).

My cousin said, "When I was at our uncle's, he did something much more interesting. As before, he divided up the dimes for ice cream by taking a number, multiplying it by itself, and distributing the dimes by using the number he got. As before, he gave the smallest child as many dimes as the first digit (reading from left to right). The next smallest child got as many dimes as the second digit, and so on. Since I was the biggest, I got as many dimes as the last digit.

"But the dimes were distributed more fairly than when you visited him. The four largest children—I among them—got the same number of dimes, and so we did not quarrel."

At this point, my cousin paused to take a breath, and I interrupted:

"So if this is how it happened, you couldn't have bought yourself any ice cream."

Was I right?

Hint: p. 78 Solution: p. 146

24th Week

A Ship Sails By

Several children are arguing on the river bank. A ship is sailing on the river, parallel to the shoreline. The children would like to measure the length of the ship. Some of them think it can't be done since they don't know the speed of the ship. But Tommy has already started working on it. He reckons the length of the ship is 200 paces, if we are walking in the same direction as the ship. If we walk in the opposite direction, however, the ship's entire length passes us after just 40 paces.

On the basis of these measurements, he easily figured out the length of the ship.

How can the ship's length be determined?

Hint: p. 79 Solution: p. 147

At Lake Michigan II

Tommy has laid out the matches in small piles: a pile of one, a pile of two, and a pile of three. (See 23rd Week, rules of the puzzle "At Lake Michigan I," p. 33.)

Steve goes first.

Who has a winning strategy?

Solution: p. 148

Dinner Guests

A teacher couple (husband and wife, both teachers) invited a doctor couple and an engineer couple to dinner. The blonde wife is busy in the kitchen; her husband arranges a chess board. The doorbell rings. The host asks two brown-haired and two black-haired guests to come in. After a few minutes, Stephen, Thomas, and Paul turn their attention to chess problems, while Dora, Ella, and Petra talk about their jobs.

We apologize for interrupting, and direct a few questions to those present. We ask

1. Paul, what the color of his wife's hair is (answer: brown);
2. the male engineer, what the first name of Dora's husband is (answer: Thomas);
3. Petra, whether there are two women in the party with the same hair color (answer: yes);
4. the male teacher, who his wife is (answer: Petra);
5. a black-haired member of the party, what the profession of the spouse is (answer: doctor).

We bid all farewell, and a red-haired member of the party sees us out.

Unfortunately, the value of these answers is not clear this evening because—just for the fun of it—the doctors and another member of the party never told the truth. We know, however, that the others told only the truth, so we can still determine the hair color, profession, and the name of the spouse for all present.

Solution: p. 148

25th Week

Beside/Below

The *Beside/Below* game is played as follows: We take a deck of cards. Then we lay the top card on the table, face up, move the next card to the bottom of the deck, lay the third card on the table beside the first, move the

following card to the bottom of the deck, and so on, until no card is left in our hand.

A friend showed us the following trick. He took a deck of 13 cards, and following the rules of *Beside/Below*, he placed the cards on the table in the order: Ace, King, Queen, Jack, 10, 9, 8, 7, 6, 5, 4, 3, 2.

What was the order of the cards in his deck?

Solution: p. 150

Lesson in Economy

Here is an old English story from the time when shilling coins were still in use (20 shillings equaled 1 pound):

"So," said the shipping director to his deputy, "as you know, we need to transport this shipment 1,000 miles. What will it cost us to ship?"

"That depends to a large extent on the speed of delivery," came the answer. "The base price is 1 pound per mile, with a surcharge of one shilling per mile for each mile an hour above 10. So if the truck travels at 20 miles per hour, we'll have to pay 1 pound 10 shillings for each mile of the route."

"Then it will be the cheapest if we drive at 10 miles per hour?"

"Not quite. We also have to pay a 25 pound penalty for each hour over 20."

What is the delivery truck's most economical speed?

Hint: p. 79 Solution: p. 151

Auto Racing

Cambridge is two miles from Northy, connected by a winding road. The first mile of the road climbs up to the top of a mountain, at whose foot lies Northy; the second mile winds its way down to Northy.

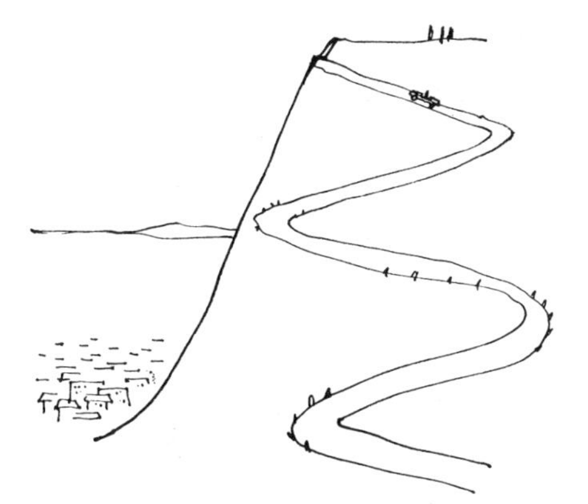

A celebrated race car driver took up the challenge, despite the extraordinary difficulty of the terrain, to drive from Cambridge to Northy at an average speed of 30 mph. Observers at the summit confirmed that on the way up, he was able to drive at a speed of only 15 mph. Nevertheless, he arrived in Northy only a few seconds late. He was buried the following day.

How did he die?

Hint: p. 79 Solution: p. 153

26th Week

Uncle Pops a Quiz

An uncle is chatting with his nephew:

"Milo, can you tell me how many equations you need to find six unknowns?"

"Six."

"OK. Now I will give you five equations and I still hope you can find the solution.

"There are six groups of workers in my factory under my supervision. I multiplied the number of workers in each group by the total number of workers in the remaining groups and I obtained six numbers. Of these six numbers, here are five: 1,000, 825, 549, 325, and 264.

"Can you figure out how many workers are in each group?"

"Well, that's not really a difficult problem; I'll try to solve it soon. It is not surprising that fewer than six equations will do; from the nature of the problem, it is clear that the unknowns have to be natural numbers."[5]

If the solution of the problem seems easy to Milo, we hope it will be no big deal for us either.

Hint: p. 79 Solution: p. 154

The Nephew Strikes Back

"Believe it or not, I've already solved the puzzle," says the nephew after a short while, "and I've noticed something interesting (see the previous puzzle, "Uncle Pops a Quiz," p. 38). If you had asked how many workers work in your factory, then two numbers would have sufficed, for instance 1,000 and 264."

"That's remarkable," says the uncle, astonished. "Why is that true?"

[5]For the basic terminology, see Section 1 of the Appendix, p. 221

Let's give the uncle a hand, so he will not be embarrassed in front of his nephew.

Solution: p. 155

The Window Dresser's Error

My girlfriend Judith is a window dresser. She related the following incident:

Judith arranges digits written on small pieces of cardboard to form the prices displayed in the shop window. On one occasion, she had just placed four of these small cards next to a hat, when the manager told her that the price she had displayed was incorrect; the actual price was a multiple of the one she had given. The error was easy to correct, however. All she had to do was to reverse the order of the cards.

Can you figure out the price of the hat if I tell you that the displayed price and the actual price are both square numbers?

Solution: p. 155

27th Week

At Lake Michigan III

Tommy and Steve played several games with just two piles of matches. But soon they had discovered the following simple rule: if played correctly, with an equal number of matches in the two piles, the player who makes the first move always loses; with an unequal number of matches in the two piles, the player who makes the first move always wins. (See the rules of the puzzle "At Lake Michigan I" (23rd Week, p. 33).)

How do you win in the two cases?

Solution: p. 156

Colored Dice

At math camp, the children were painting dice with six different colors. As usual, the faces of the dice were numbered from 1 to 6 so that the

numbers 1 and 6, 2 and 5, and 3 and 4 were on opposite faces of the dice. The children painted the six faces of the dice with six different colors. (All of the children used the same colors.)

Each child painted 20 dice, and among the painted dice, no two were the same with respect to color and numbering.

What can we say about the number of children at the camp?

Hint: p. 79 Solution: p. 156

Francesca's Teachers

In Francesca's grade, the subjects mathematics, physics, chemistry, biology, English, and history are taught by three teachers, Ms. Anderson, Mr. Knight, and Ms. Evens.

(1) Each teacher teaches exactly two subjects.

(2) Ms. Anderson is the youngest of the three teachers.

(3) The math teacher and Ms. Evens always go swimming together.

(4) The chemistry teacher lives in the same house as the math teacher.

(5) The physics teacher is younger than Mr. Knight; the biology teacher, on the other hand, is younger than the physics teacher.

(6) The eldest of the three teachers lives alone nearest to the school.

Can you figure out who teaches which subjects?

Solution: p. 157

28th Week

Scribble

"You've scribbled all over the back of an important letter again."

"Don't be angry, Papa, but since I worked out that puzzle—how to draw a little house without lifting the pencil or drawing a line twice ("Nicholas's House," 3rd Week, p. 4)—I'm forever drawing figures like that, with a single, unbroken pencil line."

"And this is supposed to be such a figure?"

"Yes."

Even though the father made several attempts, he didn't succeed in duplicating his son's achievement. Maybe we'll have better luck.

Solution: p. 157

Randolini's Card Trick

Randolini, the famous magician, never failed to astonish his audience with the following card trick.

He hands a volunteer from the audience a deck of cards. Then he turns, so his back is to the volunteer, and instructs her to carry out the following steps:

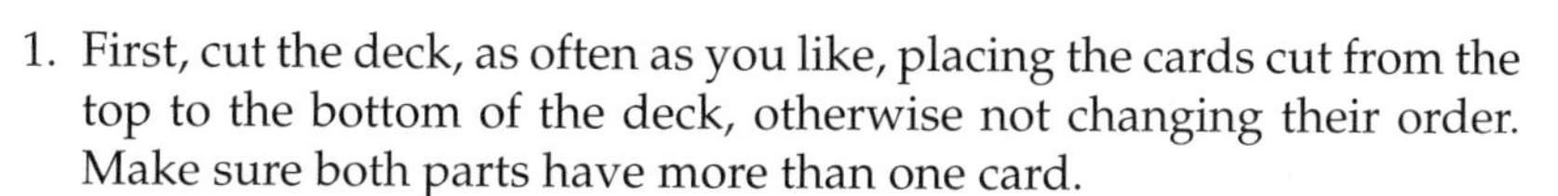

1. First, cut the deck, as often as you like, placing the cards cut from the top to the bottom of the deck, otherwise not changing their order. Make sure both parts have more than one card.

2. Then, cut the deck and place the top part next to the bottom part. (There are now two stacks of cards.)

3. Next, pick up one of the stacks and slip its bottom card anywhere into the other stack. Remember this card!

4. Finally, shuffle one of the stacks and hand it to me.

Randolini doesn't even ask which stack he is getting: the stack from which the volunteer removed a card, or the stack to which she added the card.

Then Randolini holds in his hand the stack that was returned to him. He ponders for a moment, looks hard at the cards, and then names the card that was moved from one stack to the other. And if this card is in the stack Randolini holds, he shows it to the applauding audience.

How does Randolini's trick work?

Hint: p. 80

Solution: p. 158

Vacation

I have five good friends, Dallas, Orlando, Memphis, Jackson, and Phoenix. My friends live in cities named like them, but none lives in the city that bears his own name.

This year, each of them spent his vacation in one of these cities. We know that no two of my friends live in the same city or went on vacation in the same city, that none spent his vacation in the city bearing his name, and that none spent his vacation in the city he lives in.

From my friends' letters, I gathered a few other facts:

Jackson traveled to Phoenix. Orlando spent his vacation in the city that is named like my friend who lives in Phoenix. My friend who lives in Dallas spent his vacation in Jackson, whereas my friend who lives in Phoenix spent his vacation in the city that is named like my friend who lives in Dallas.

Who traveled to Orlando? Can we find out in which city my friend Phoenix lives?

Solution: p. 158

29th Week

A Typical Mathematician

In response to being asked his age, Augustus De Morgan, the great English mathematician of the nineteenth century, gave the following answer: "I was born in this century, and was in the fortunate position that in the year x^2 I was just x years old."

When was De Morgan born?

Solution: p. 160

Weight Trick II

Another intriguing problem (we met the first in the 18th Week as "Weight Trick I" on p. 25) is the following: eight of nine boxes weigh exactly the same; one of the nine is somewhat lighter than the others.

How many times do you have to use the balance scale in order to determine which is the lighter box?

Solution: p. 160

At Lake Michigan IV

Working diligently, Steve and Tommy figured out a winning strategy for any number of piles. They named it the *Binary Strategy*. We let Tommy explain:

"Let's write the number of matches in the piles *as binary numbers*. We'll write the binary numbers under each other (aligned on the right) and add up the columns (without carries) *as decimal numbers*.

"We have a winning position—call it the Binary Winning Position—if we obtain an even number in each column and it is our opponent's turn. If there is an odd number in any column, and it is our turn, we have to play in such a way (and we always can!) that after our move, we get an even number in every column, and thereby get into a Binary Winning Position. We continue playing in this way until we get into a One-Peek Position (with one exception, there is only one match in each pile; see 23rd Week, "At Lake Michigan I," p. 33), so we know how to win."

Try out this winning strategy in the previous examples! Consider also a new example in which there are four piles: one pile of six matches, two piles of five matches each, and one pile of three matches.

Solution: p. 161

30th Week

Carton Puzzles

How many different ways can you read CARTON PUZZLES in the illustration?

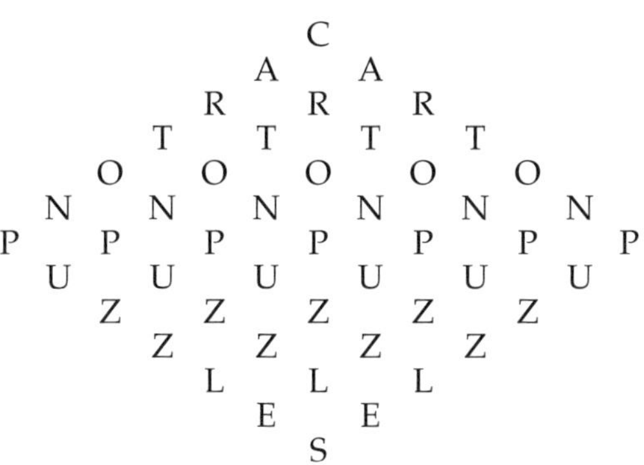

Start from the letter C at the top, proceed diagonally down to either the right or the left, and finish with the letter S at the bottom.

Hint: p. 80 Solution: p. 162

No Splitting of Hairs!

Are there two people in Salt Lake City who have exactly the same number of hairs? Is it possible that there are even more such people? If so, how

many are there? (A full head of hair is about 100,000–150,000 hairs, and Salt Lake City has a population of about 1,700,000.)

Solution: p. 164

Toy Soldiers—Halt!

Little Robert is playing with his toy soldiers. He sets them up so that each row has the same number of soldiers. Then he goes to his father to tell him what he has done.

"Five of my soldiers were sick, and I set up the rest in 24 even rows. Then all the soldiers were well again, and I sent 16 soldiers off to scout the forest to see if the enemy was there. In the meantime, I had the bugler call the others to assembly, and they lined up into 15 even rows."

"Well now, Robert, you've made a tiny error here. The rows couldn't have been even."

Why not?

Hint: p. 80 Solution: p. 165

31st Week

Dicey Question

I ask a friend to throw three dice—colored red, blue, and green—while my back is turned, and note the values of each dice, by color. (The uppermost face of a die shows the *value* of a throw.)

Then I ask him to multiply the value of the red die by 2, add 5 to the result, then multiply the number he gets by 5. To this number he should add the value of the blue die, multiply the sum by 10, and add to the result the value of the green die. Then he should tell me the final result.

On one throw, for example, the final result is 484. I tell him that the three values of the dice (in the order red, blue, green) are 2, 3, and 4.

How do I figure this out?

Solution: p. 166

Party Question

At a party, we count the number of friends of each party goer. We do not assume that all present are friends. Friendship is mutual (if one person is friends with another, then the latter is friends with the former).

We find that at this party an even number of people have an odd number of friends.

Is this true at every party?

Hint: p. 80 Solution: p. 167

First and Last Day of the Year

In 2010, January 1st and December 31st both fall on a Friday. In which other years is it true that the 1st of January and the 31st of December fall on the same day of the week?

Hint: p. 81 Solution: p. 169

32nd Week

Bridge Party

On a winter evening, Misters North, East, South, and West are playing bridge. The card table is positioned so that the players are seated on the

North, East, South, and West sides of the table (indicated in the illustration by the letters N, E, S, and W, respectively, on the table).

After each rubber (a *rubber* is a series of games in which the winning player reaches a certain number of points), the players draw to determine on which side of the table they will play and with whom they will make up a partnership (partners face each other across the table and play together against the other two players). During no rubber on this particular evening did any player sit in the same place as his name. The seating order was different for every rubber, and in every rubber Mr. West was a member of the victorious partnership.

Mr. North later claimed that he won three rubbers. Was he telling the truth?

Hint: p. 81 Solution: p. 169

Beside/Below Redux

I related the *Beside/Below* game (25th Week, "Beside/Below," p. 36) to George, and he liked it so much that he thought up a bunch of card sequences and assembled the corresponding decks. So, for example, he assembled a gigantic deck of 971 cards, numbered from 1 to 971.

He showed his sister the new deck of cards and told her that the cards were arranged in order from 1 to 971, the card with number 971 at the bottom of the deck. Then, according to the *Beside/Below* method, he began laying out the cards. As he laid them out, he asked his sister:

"What do you think will be the last card left in my hand?" Then he asked, "Do you know in which round the card 228 will be laid on the table? And which card will be laid out in the 634th place?"

How did George's sister answer the questions?

Hint: p. 81 Solution: p. 170

Weight Trick III

We have 12 boxes and know that 11 of them weigh the same. Using the balance scale three times, can we determine which of the boxes is of a different weight and whether it is lighter or heavier than the others?

Hint: p. 81 Solution: p. 171

33rd Week

The Spy Who Came around the Corner

The following map shows the office buildings in a huge factory complex.

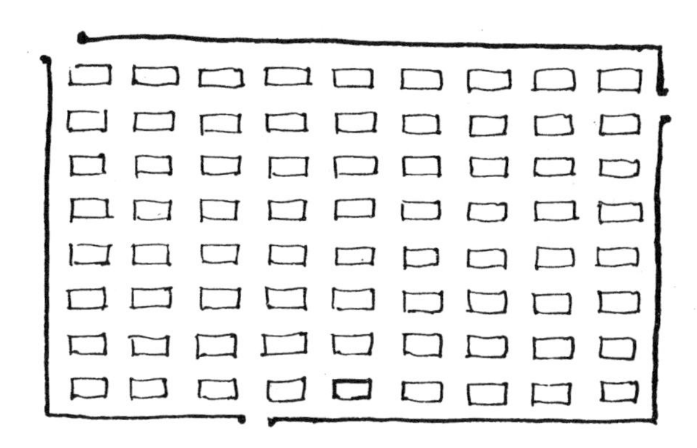

For security reasons, a high wall was built around the complex, containing a total of three gates, shown on the map.

A spy working in the factory complex smuggled out important information from time to time. It seemed difficult to find the spy, since he was very clever. Once, however, the spy made a fatal error, and they intercepted a message he sent to his contact.

Here is his message:

> We'll meet every Tuesday on the corner next to my office building. Come through the gate that is on the upper-left corner of the map. To not make it obvious that you often make visits here, please take a different route each time. That is easily done, since there are 715 different routes available to you.

We assume, of course, that the visitor made no detours—that is, all routes were as short as possible. Based on this information, the meeting place was quickly located and the spy was caught.

On which corner was the meeting place?

Hint: p. 81 Solution: p. 172

Grandfather and Grandson

There is a story about a politician who in 1932 was exactly as old as the number formed by the two last digits of his birth year. What caught our attention is the fact that the same was true of the grandfather of that person.

How is this possible?

Solution: p. 173

Finally, Concerning Jealous Husbands

Five jealous husbands and their wives came to a river and wanted to cross. They were aware of the previous problems (7th Week, "Jealous Husbands," p. 10; 12th Week, "More Jealous Husbands," p. 17; and 21st Week, "Still More Jealous Husbands," p. 30). This time, they had available to them a three-person boat and they successfully crossed the river.

How did they manage it?

Solution: p. 173

34th Week

One More Time: Knight on the Chess Board

In the solution of "Knight on the Chess Board" (18th Week, p. 26 and p. 131), we saw that we cannot move from the lower-left corner to the upper-right corner of a chess board in an odd number of moves such that with each move we land on a square of the opposite color. Thus, it is not possible for a knight, starting in the lower-left corner, to move across the chess board so that it lands on each square exactly once and ends up in the upper-right corner.

Now, omitting the condition that it must start in the lower left corner and finish in the upper right corner, let's see if we can move the knight across the chess board so that it lands on each square exactly once.

Solution: p. 174

Battle of Cards

Andy, Bert, Chris, and Dan are playing cards. After one round, Andy, Bert, and Chris have twice as much money as they started with. In the second round, Andy, Bert, and Dan double their money. In the third round, Bert loses so much that Andy, Chris, and Dan double their money. Finally, in the fourth round, Andy loses so much that Bert, Chris, and Dan double their money. After the fourth round, they each have $64.

How much money did the players start with?

Hint: p. 81 Solution: p. 175

Remains of a Long Division

In one of my desk drawers, I found a faded sheet of paper on which a division problem had once been written. The digits shown below are those I could decipher with certainty (the illegible numbers are indicated by the short lines):

```
            5__3
  ___ |1____5_
        __9_
        ====
          ___
          __6
          ===
          ____
          ====
           __3_
           ====
              0
```

What was the division problem?

Hint: p. 82 Solution: p. 177

35th Week

Ten Little Slips of Paper

We write the numbers 1 through 10 on ten slips of paper, and we put them into a hat. Anna, Greg, Louis, Harriet, and Maria each draw two slips of paper from the hat. The sums of the two numbers drawn by each child are 17 (Maria), 16 (Harriet), 11 (Anna), 4 (Greg), and 7 (Louis).

Who drew which slips of paper?

Solution: p. 179

Rook Walk

The rook stands on the lower-left corner of the chess board. Can it be moved to the upper-right corner so that it crosses each square exactly once along the way.

Hint: p. 82 Solution: p. 180

Married Couples

Three men (Andy, Bert, and Chris) went with their wives (Anna, Bella, and Crystal) to a discount store. When they finished shopping, each of the six realized that the average price (in dollars) of all the items he or she had individually purchased equaled the number of items each had bought.

Andy bought 23 more items than Bella. The average price of the items Bert bought was $11 higher than the average price of the items Anna bought. Each husband spent $63 more than his wife.

What is Andy's wife's name?

Hint: p. 82 Solution: p. 180

36th Week

Legal Problem from Antiquity

Caius and Sempronius arranged a feast. Caius brought seven dishes and Sempronius eight. Titus arrived as an unexpected guest, and the three diners divided the dishes among themselves. Titus ate 30 dinar's worth of food, and so he said:

"The ratio of dishes you furnished is 7 : 8, so I'll split my 30 dinars between the two of you accordingly."

So, Titus gave Caius 14 dinars and Sempronius 16 dinars.

Sempronius objected to this. His fellow diners paid him no attention, so he submitted his problem to the court.

What was the judgment of the court?

Hint: p. 82 Solution: p. 181

At the School Dance

At a school dance, 430 different couples danced. Girl 1 danced with 12 different boys, girl 2 with 13 different boys, and so on, and the last girl danced with all the boys.

How many girls and how many boys attended the dance?

Solution: p. 181

Square Numbers

The numbers 190,246,849 and 190,302,025 are the squares of two consecutive odd numbers.

Without the use of a calculator or a computer, figure out the square of the even number that lies between the two odd numbers.

Hint: p. 82 Solution: p. 182

Black Belt

37th Week

Half-Truths

Finally, the five friends—Kate, Eve, Susie, Paula, and Rose—finished the table tennis tournament. They decided to play a trick on their parents: each of them would tell her own parents only a half-truth, that is, one true and one untrue statement. The statements were the following:

Kate: Paula placed second in the tournament. Unfortunately, I was only third.

Eve: I finished first, but Susie finished second.

Susie: I got third place; poor Eve, on the other hand, was last.

Paula: I finished second, while Rose finished fourth.

Rose: I got fourth. Kate was lucky; she got first place.

Can we determine the truth?

Solution: p. 182

The Great Hunt

Four hunters, Andy, Bert, Chris, and Dan, bagged only four different kinds of animals: wild boars, deer, wolves, and foxes (at least one of each). They agreed to a contest in which the bagged game, in the order *boar, deer,*

wolf, fox, were assigned points in a decreasing sequence of natural numbers. Among the game killed were three wolves, but only one wild boar, which was shot by Chris.

The four hunters won a total of 18 points. Dan had the lowest point score, although he had bagged the most game. Andy and Dan together had as many points as Bert and Chris together.

For each hunter, list the animals he shot.

Solution: p. 183

38th Week

Headscratcher with Caps

I ask three friends, all mathematicians, to seat themselves, one behind the other, to act out a puzzle.

I tell them I will blindfold them and put caps on their heads, selected from two yellow and three green caps. Then I remove the blindfolds, and I suggest that each of them should try to determine the color of the cap on his head. However, no one is allowed to look behind, so each person can only see the head(s) of the person(s) in front of him.

Shortly after I remove the blindfolds, the person sitting furthest back says, "I can't tell the color of the cap on my head."

Then the person sitting in the middle says, "I can't tell the color of the cap on my head, either."

What did my friend sitting up front say?

Hint: p. 82 Solution: S. 185

Schoolgirls

Ten girls, Anna, Betty, Crystal, Doris, Edith, Frances, Giselle, Heidi, Irene, and Jessica, attend a private school in New York City, where the students range from Grade 1 to Grade 12. They are in ten different grades, and no girl is in a grade higher than tenth.

Here is what we know about the girls:

1. One girl is Edith's cousin and is in the seventh grade; another is Frances's cousin and is in the fifth grade.

2. Anna is one grade above Irene; Frances is two grades below Jessica.

3. This year, Crystal is not yet in tenth grade.

4. At the conclusion of the third grade, Heidi made the honor list.

5. Crystal has already completed the fourth grade, Betty the fifth.

6. Anna, Irene, and the sixth grader belong to the biology club of the lower school; Betty, Frances, and the eighth grader belong to the biology club of the upper school. (The lower school is the first six grades in this private school.)

7. Anna and the girl from the seventh grade live in the Garment District; Doris and the fifth grader live in Chelsea; Edith, the first grader, and the girl from the eighth grade live in the West Village; Irene and the girl from tenth grade live in Scarsdale, a town north of New York City.

8. Girls behind in their studies are tutored by students from a higher grade: Betty tutors Frances, Edith tutors Heidi, and Doris tutors Anna.

List the girls and the grades they are in.

Is every one of these statements required to solve the puzzle?

Hint: p. 83 Solution: p. 185

39th Week

Tennis Tournament

I wrote a news article about a round robin tennis tournament—each player played every other player once. (Recall that in tennis there are no ties.)

I thought it would be interesting to pick a player and ask which players he defeated. Then I would also ask all those he defeated the same question. I knew that any player not mentioned would be offended.

Is it possible to chose a tournament player so that every participant would be in the article?

Hint: p. 83 Solution: p. 187

Hurdle Race

A hurdle race was organized among five teams of the same size. The winning team would be awarded $10 gift certificates; every runner on that team receives at least one gift certificate, and a runner placing higher in the race would receive more than someone placing lower; moreover, no two runners get the same number of certificates.

The organizers declared that each team must have different rules by which the gift certificates would be distributed among its members; different rules yielding different distributions.

A member of the winning team is a good friend of mine. He told me the following about the awards—and meant thereby to confront me with a pretty brain teaser:

"As it happened, the organizers had thought carefully about how large the teams should be and about how many gift certificates to distribute. The number of gift certificates was divisible by 5 and the gift certificates could be distributed in exactly five ways. Our team won and I placed third in the team; together with the fourth-place runner, we received exactly as many certificates as the first-place runner."

Then he described another similar race:

"Later in the day, and under the same rules, a second race took place among five equal-sized teams of runners. The number of gift certificates distributed as well as the size of the teams were different, but there too, in accordance with the rules, the gift certificates could be distributed in exactly five ways, and the number of gift certificates was divisible by 5. The first-place runner on the winning team in the second race received only as much as the second-place runner in the first race. But in the second race also, the third-place and the fourth-place runners together got the same number of gift certificates as the first-place runner. In addition, in both contests, fewer than 60 gift certificates were given out."

Then he asked me if I could figure out how many gift certificates he had received. Could I do that?

Hint: p. 83 Solution: p. 188

40th Week

Vacation Days

At year's end, everyone in my company is entitled to 20 days of vacation, which can be taken in December and January. This year, the company mathematician noticed an interesting fact: a number of people took the 20 vacation days in three segments, moreover, each in a different grouping of days. In addition, the mathematician found that, as a result, all possibilities were exhausted, including those cases in which the number of days taken in the three blocks were the same, only in a different order. The number of these employees was exactly 15% of the entire work force.

How many employees does my company have?

Hint: p. 83 Solution: p. 191

The Wine Grower's Estate

A wine grower bequeathed 45 barrels to his cousins, Andy, Bert, Chris, Dan, and Eddy. One fifth of the barrels were full of wine, one fifth were three-quarters full, one fifth half full, one fifth a quarter full, and one fifth empty.

The five cousins divided the estate so that each of them received the same number of barrels and the same quantity of wine. They managed it so cleverly that not once did wine have to be transferred from one barrel to another.

Each cousin received at least one barrel of each kind; no two cousins received the same number of every kind of barrel, and the number of full barrels was never the same as the number of three-quarters full barrels. Andy got the most empty barrels, Bert and Dan got the most quarter-full barrels, while Dan and Eddy got the fewest full barrels.

How did the cousins divide the barrels among themselves?

Solution: p. 192

41st Week

Divine Injunction

Three diamond needles, attached to a marble table, glitter beneath the dome of an ancient temple. Long ago, when the temple was built, there were 64 perforated golden disks on the first needle, the largest on the bottom, a smaller one next, on top of this an even smaller one, and so on, so that the disks were arranged by decreasing diameters.

The divine injunction was this:

"All 64 discs have to be transferred to the second needle. The transfer shall be accomplished utilizing the third needle. Follow these instructions:

1. Only *one* disk may be moved at a time, and only the uppermost.

2. A disk may be transferred only to an empty needle or to one on which the diameter of the uppermost disk is larger than that of the disk being transferred.

3. One transfer must occur every second.

When the last disk is transferred, the world will come to an end."

In about how many years will the world come to an end?

How many moves are needed to complete the transfer?

Hint: p. 83 Solution: p. 193

Mathematicians in Conversation

Andy, Bert, Chris, and Peter are having a good time together. All four are professional mathematicians. One of them is notoriously absent-minded.

Peter asks the other three how old they are. They agree to give their ages in a number system other than decimal. Here are their answers:

"Let us consider four one-digit numbers that succeed each other by a difference of two," says Andy. "After the smallest number, write the next one, and the two-digit number you get gives my age, which is the same as the product of the two largest numbers of the four."

It is Bert's turn next: "We get my age if we write two consecutive one-digit numbers together, forming a two-digit number that equals the product of the two numbers following those we have selected."

"I am the youngest among us," says Chris, "and I can make exactly the same statement about my age as Andy made about his."

Then Peter figured out the ages of Andy, Bert, and Chris, and who the absent-minded mathematician was, who, true to form, acted somewhat absent-mindedly.

Can we succeed as Peter did?

Hint: p. 84 Solution: p. 195

42nd Week

Gifts

Mr. Smith has five daughters: Eve, Frances, Clara, Mary, and Susie. Mr. Smith tells his daughters, "Your four cousins, Bert, Josh, Robert, and Simon, are coming to visit tomorrow. I'm giving you each $80; please buy some gifts for the boys. To keep you on your toes, I ask you to observe these conditions:

"Each of you should buy something for each boy, and the price of each gift should be an integer multiple of ten.

"Make sure you each divide your $80 among the four boys differently; there should be no two equal divisions, not even in a different order (that is, if one girl gives the cousins gifts valued at $10, $10, $10, $50, respectively, then another girl cannot give $50, $10, $10, $10). To make sure that none of the boys feels slighted, all four should receive gifts of an equal total value. And, of course, each of you must spend the entire $80."

The girls stick to their father's stipulations.

Frances buys more for Bert than for the other three boys combined.

Clara spends on Simon and Robert together exactly as much as Frances spends on the other two.

Mary spends more on Josh than on the others; Eve does the same for Robert.

How do the girls split up the money they received for the gifts?

Hint: p. 84 Solution: p 196

The Fickle Sultan

The sultan locks one hundred prisoners into one hundred cells, one prisoner in each cell. The cells have locks with two positions: when turned, they alternately unlock and lock the cell doors. The prisoners cannot see or hear when their doors are locked or unlocked.

After the prisoners are locked into their cells, the Sultan changes his mind and sends a guard to run down the line of cells giving one turn to every lock. Then he changes his mind again, and quickly dispatches a second guard to give every second lock a turn. No sooner has he done this, he sends yet a third guard with the order to give a turn to every third lock. In this manner the Sultan proceeds until finally he gives the hundredth guard the order to give one turn to every hundredth lock.

Now he decrees that every prisoner whose cell is unlocked can go free. Which cell doors are unlocked at the end?

Hint: p. 84 Solution: p. 198

43rd Week

Military Band—Blow Your Own Horn!

The military band has five members: Burnes, Davis, Matthews, Patterson, and Vance. Each has a different rank: the band consists of (in ascending order) a lieutenant, a captain, a major, a lieutenant colonel, and a colonel.

Each of them has just one sister and no brother. All five married the sister of one of the others. None of the officers' wives, however, is the sister of the husband of the sister of her husband. At least one of Matthews's brothers-in-law carries a higher rank than Matthews himself.

Both of Davis's brothers-in-law went with the military band touring in France, as did Patterson's two brothers-in-law. But none of the colonel's brothers-in-law went along on that tour.

The captain has never visited Japan.

Patterson was together with his two brothers-in-law at a festival in Finland; the lieutenant, on the other hand, has never been to Finland.

The lieutenant colonel was together with his two brothers-in-law on a tour in Canada.

Patterson and one of his brothers-in-law have never been to Canada.

The colonel was together with his two brothers-in-law on a tour in Japan, but has not yet been to Canada.

Vance has never been to Japan or Finland.

List the band members, by name and rank.

Hint: p. 84 Solution: p. 198

At the Round Table

Each weekday at noon, four married couples eat lunch at the same round restaurant table. They always arrange themselves at the table so that no couple sits next to each other.

After meeting this way at lunch for a month, one member of the group asked:

"How long do you suppose we can keep seating ourselves in different arrangements, so that couples don't sit next to each other?"

There was a rather wide range of opinion. And there was also the opinion that, if necessary, they could find new possibilities for several weeks yet.

What is the answer?

Hint: p. 85 Solution: p. 199

44th Week

WaterMonster

Water flows into a circular fountain from 366 nozzles. The valve for each nozzle has two settings: in one position, no water flows; in the other position, water sprays from the nozzle. In leap years, 366 nozzles are used, otherwise only 365 (in each year, as many valves as there are days).

The valves are opened and closed by complicated electric circuits controlled by a computer program, called WaterMonster. At the beginning of New Year's Day, exactly at midnight, WaterMonster opens the valves (366 or 365, depending on whether it's a leap year). At the start of the second day (January 2), WaterMonster switches the position of every second valve (that is, the valves numbered 2, 4, 6, ..., counting clockwise from the ornamental statue on the fountain). At the start of the third day, WaterMonster switches the position of every third valve (that is, the valves numbered 3, 6, 9, ..., again counted clockwise from the statue), on the fourth day every fourth valve, and so on, up to the 365th day (in non-leap years) or the 366th day (in leap years).

On the 31st of December, how many (and which) nozzles spray water?

Solution: p. 202

Tommy's Surprise

Tommy—who knows how fond I am of multiplication puzzles in which only a few digits are visible—surprised me on my birthday with this puzzle:

```
            . . T . . . T
          ×   . . . . T .
          ---------------
          . T . . . . . .
        . . . . . . . .
        . . T . . . T
    T T . . T . . T
  . . T T T . T .
. . . T . . . .
-------------------------
T . . T . . . T . . . . .
```

He added the following comments:

"The letter T always represents the same digit, and no period stands for the digit represented by the letter T. It's not a difficult puzzle, and if you really apply yourself, you may finish by the time your next birthday comes around."

Well, it didn't actually take me that long. I hope it won't take you so long either.

Hint: p. 85 Solution: p. 203

45th Week

A Number in Mind

Alfred has a positive natural number in mind and tells us this much about it:

"The square of the number larger by one differs only in the order of its tens digit and its ones digit from the square of the number I have in mind."

Can you guess his number?

Hint: p. 86 Solution: p. 204

The Hungry Sales Woman

A woman selling fruit at the market has three baskets of apples in front of her. She sells an apple from a basket for as many cents as there are apples in the basket in the morning. If we multiply the money she made from the sale of the apples from the first basket with the money she made from the second basket, the result is the same as the amount made from the sale of all three baskets.

Since she can spend on her dinner only the amount she earned, she went hungry that evening. Why?

Hint: p. 86 Solution: p. 206

46th Week

Another Fragmentary Division

This is a division:

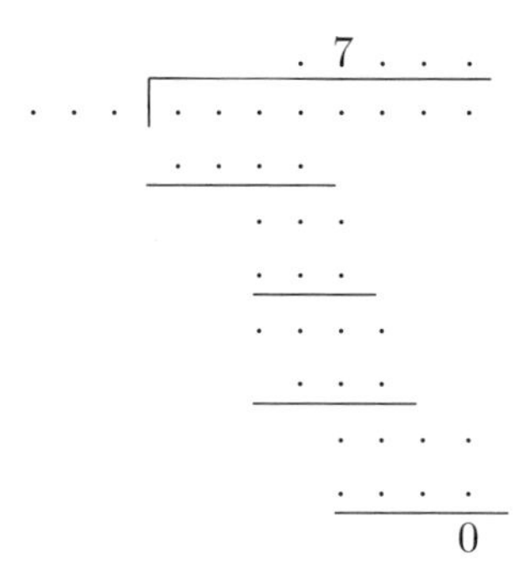

The periods stand for the digits; we reveal only two, a 0 and a 7. So little information, so many unknown digits ....

Could this conceivably be enough information to find all the missing digits?

Solution: p. 207

Return of the Sultan

This puzzle is an interesting variation on "The Fickle Sultan" (42nd Week, p. 60). The Sultan again orders one hundred prisoners locked up in one hundred cells, one prisoner per cell. (The cells have locks with two positions: open and locked. The position of the lock is changed by giving it a turn. The prisoners cannot see or hear when their doors are locked or unlocked.)

When the prisoners are locked in, the Sultan changes his mind, and sends a guard down the row of cells with the order to give each lock a turn. Then the Sultan changes his mind again and sends a second guard to give every second lock two turns. No sooner has he given this order, he sends yet a third guard to give every third lock three turns. He continues this pattern until he finally issues the order to the hundredth guard to give every hundredth lock one hundred turns.

Which cell doors are unlocked at the end?

Hint: p. 86 Solution: p. 208

47th Week

An Interesting Game

This is a game for two people. Players take turns laying down quarters on a rectangular table until no more coins fit on the table top. Coins may touch, but may not overlap each other. The winner is the player who places the last coin on the table.

Is there a winning strategy for the player who goes first?

Hint: p. 86 Solution: p. 209

At Lake Michigan V

Is the Binary Strategy (which Steve described in the puzzle "At Lake Michigan IV" (29th Week, p. 42)) a winning strategy?

Solution: p. 209

48th Week

Computational Wizard

My friend George is exceptionally fast at calculating. Recently, I saw him very quickly calculate the following complicated expression. Let's do the same!

$$\frac{6 \cdot 27^{12} + 2 \cdot 81^{9}}{8{,}000{,}000^{2}} \cdot \frac{80 \cdot 32^{3} \cdot 125^{4}}{9^{19} - 729^{6}}.$$

Solution: p. 211

The Puzzling Scribble

In the solution of "Scribble" (28th Week, p. 40), we saw the following: if a zigzag drawing—which in the solution of "Party Question" (31st Week, p. 45) we called a *graph* (see p. 167)—contains more than two vertices at which an odd number of edges end, then we cannot draw this "scribble" so that we trace each line only once without lifting our pencil.

Can a graph always be drawn (in the manner described above) if it doesn't have more than two such exceptional vertices?

Of course, not. For example, the graph shown in the illustration *certainly cannot* be drawn without lifting the pencil, even though it has no vertices at which an odd number of edges end. Can you prove the claim for the graphs it would apply to?

Solution: p. 211

49th Week

Footrace

I watched a footrace, in which twelve nationally known runners started, wearing sports jerseys numbered from 1 to 12. Once the race was over, I made the intriguing observation (inclined as I am to numerology) that if the number on a contestant's jersey was multiplied by his ranking in the race, the product was always greater by 1 than a number that is divisible by 13.

Where did each runner place in the race?

Hint: p. 87 Solution: p. 213

No Tricks

Mr. Smith asks his neighbor, a retired mathematician, about the occupants of a house. As we will see, the old gentleman does not necessarily supply the most expedient answers.

"How many people live in the house?" asks Mr. Smith.

"Three."

"What are their ages?"

"I won't say, although I can tell you that 1,296 is the product of their ages."

"Well, I still can't determine their ages."

"How about this? The sum of their ages is the same as the number of the house."

Mr. Smith breaks out in a sweat as he tries to solve the puzzle, and then he says:

"I still can't figure out how old the people in the house are."

"But surely you know how old I am?" asks the old gentleman.

"Yes."

"Well, all three are younger than I am."

"Many thanks for this valuable information. Now I know how old these three people are," answers Mr. Smith.

But how should we proceed? We don't know the venerable gentleman's age, let alone the number of the house.

Is there a way to determine these ages?

Hint: p. 87 Solution: p. 213

50th Week

The Balance Scale

We want to get a set of weights for a balance scale in order to weigh all whole numbers of pounds from 1 pound to 40 pounds.

What is the smallest number of weights that will do the job?

Solution: p. 215

Unsuccessful Decree

The ruler of a land once decreed that its single richest man use his fortune to double the wealth of every other citizen. After this decree was carried out, people were amazed to discover that the size of the individual fortunes remained unchanged, although they were now in the possession of others.

How large were the various fortunes?

Solution: p. 215

51st Week

Book Lovers

"What page of the book are you on now?" Stephen asked his friend Peter.

"I won't tell you. But this much I will say: there's a prime number such that if you multiply this prime with itself many times, you'll get the page number."

"And where are you in the book?" Stephen asked, turning to Thomas.

"I'm one page further than Peter," answered Thomas. "You can figure it out by finding a prime number such that when multiplied several times by itself, it gives the page number."

"OK. Then I know which pages you two are on," said Stephen, after he had thought about it a while. "It seems you're not very far along."

Let's figure out where Peter and Thomas were in the book.

Hint: p. 87 Solution: p. 216

A Party of Eight

This party of eight people has three married couples, a bachelor, and an unmarried woman. The women's names are Judith, Maria, Margaret, and Marsha; the names of the men are Jacob, Josh, Joseph, and Michael. The men's professions are cabinetmaker, judge, locksmith, and physician.

We also know the following:

1. Joseph is Margaret's brother.
2. The bachelor met the doctor for the first time today.
3. As children, Michael and the judge were schoolmates.
4. Marsha has no relative who is a tradesman.
5. The judge's name starts with the same letter as his wife's name.
6. Josh is not related to Judith.
7. Margaret is Josh's sister.
8. Joseph has commissioned the cabinetmaker to build a wardrobe for his wife's birthday.

Husband, wife, and brother-in-law count as relatives, and of course, the relatives knew each another before the party.

Can we tell the men's professions?

Solution: p. 217

52nd Week

Weighing Once Wins!

A large factory for machine replacement parts advertised a position that required exceptional ingenuity. After some screening of the applicants, two young engineers remained in the running. The chief engineer assigned to them the following task:

"Here are 200 crates. Every crate contains 500 ball bearings, weighing one pound each. By mistake, an older type of ball bearing was packed into one of the crates. These older ball bearings each weigh exactly four ounces more than the new ones (or maybe it's the older ball bearings that are the lighter ones—I no longer remember exactly). We have a scale that can weigh thousands of pounds with precision. Can you determine with this scale, in just a single weighing—and with no other information—in which crate the old ball bearings were packed?"

With thoughtful expressions, the two engineers pondered the question. Suddenly the face of one brightened. He had hit on the solution of

the seemingly unsolvable problem. The chief engineer listened to his idea and immediately hired the young man.

Can you come up with his idea?

Hint: p. 87 Solution: p. 218

Clever Heirs

The heirs distributed their inheritance in the following manner:

The first heir received a dollars and an nth part of the remaining money. The second heir received $2a$ dollars and an nth part of the remaining money, and so on. The kth heir received ka dollars and additionally an nth part of the remaining money. Despite this peculiar method of distribution, all heirs received the same amount of money.

What is the number of the heirs?

Solution: p. 218

Hints

1st Week

■□□

A Journey by Ship

How much time passes between the departure of the first ship and the last?

Solution: p. 89

Two Games of Chess

My sister plays the two opponents at the same time.

Solution: p. 89

A Journey by Air

The wind increased the travel time.

Solution: p. 90

2nd Week

■■□□□

The Clock

We may assume that the trip to his friend's took the same amount of time as the return trip home. In what sense was setting the clock in motion important, if it doesn't give the exact time?

Calculate with variables, and the situation becomes clearer!

Solution: p. 92

3rd Week

An Unusual Year

We assign the numbers 1 to 7 to the days of the week for the whole year in some cyclical order. Assume that we assign the number 1 to the 13th. From this starting point, we compute the numbers assigned to the 13ths of the months. The puzzle can now easily be solved.

Solution: p. 93

Competitors, or *The Raise*

Calculate the paychecks from the two options for the first few months.

Solution: p. 94

4th Week

The Ancient Greek Copper Lion

Let x be the time (in hours) in which the basin is filled if water flows simultaneously from all four orifices. Find an equation to determine x.

Solution: p. 95

The Arab

Note that

$$\frac{1}{2}+\frac{1}{4}+\frac{1}{5}=\frac{19}{20}.$$

Solution: p. 96

5th Week

On the Streetcar

The passenger gave the conductor 1 forint, but not a 1-forint coin.

Solution: p. 96

Hunting Hares

Let x be the number of leaps the hare makes before the greyhound catches up. Find an equation for x.

Solution: p. 96

Telephone Cable—A Long Line?

The circumference of a circle with radius r is $2\pi r$, where π is approximately 3.14.

Solution: p. 97

8th Week

Another Superhotel

Has the question been framed correctly? Why should the addition in question yield a sum of $300?

Solution: p. 103

Three-Letter Words

Start with $N + R = N$ and note that some steps of the additions result in a digit to be carried and some do not.

Solution: p. 104

In the Kindergarten

Arrange the children into four groups so that children whose eye and hair colors match are in the same group.

Solution: p. 105

9th Week

How Old Is Susie?

A *prime number* is a natural number that is divisible (without a remainder) only by two numbers: 1 and itself. The number 1 is not considered a prime number. For example, 6 is not a prime number because it is divisible by 2 and 3; the number 7, on the other hand, is a prime number; the number

2 is the only even prime number. (There is a table of prime numbers in Section 4 of the Appendix; see p. 225.)

Solution: p. 106

Curious Addition

The addition is not carried out in the decimal system.

Solution: p. 106

10th Week

The Local Lurches On

Let x be the distance the local train travels with the technical defect before the express train catches up with it. Find an equation for x.

Solution: p. 111

Cocoa Island

What answer can a native give to the question "Are you Bodo or Wizzer?"

Solution: p. 112

The Little Troublemaker

Remember: identical chess pieces don't necessarily cover identical digits.

First, figure out the quotient; the digits of the quotient can quickly be derived from the lengths of the subproducts.

Solution: p. 112

11th Week

Five Little Decks of Cards

The second time we lay out the cards, we arrange them so that in every deck there is only one card from the original corresponding small deck, and indeed in an order known to us.

Solution: p. 113

A Clumsy Division

Here, too, first determine the quotient.

Solution: p. 113

12th Week

Cycle Tour I

Going backwards from 100, we search for the positions in which winning or losing is certain. Proceeding this way, we can answer the question.

Solution: p. 114

Toy Soldiers—On the Double!

Use the fact that each integer n can be written in the form $n = 36k + r$, where r is a nonnegative integer smaller than 36. We say the n divided by 36 is k with a *remainder* r.

Solution: p. 115

13th Week

Big Problem—Chess Problem

If we place one rook on a square, how many other squares are available for the other rook?

Solution: p. 116

Cycle Tour II

Go through each case of a limit (5, 6, 7, 8, or 9) systematically, according to the idea of the solution of the puzzle "Cycle Tour I" (12th Week, p. 16).

Solution: p. 116

14th Week

The Barber

Can the barber shave himself?

Solution: p. 120

15th Week

Car Trip

Let x be the distance from the point at which the car passes the truck to Chicago. Find an equation for x.

Solution: p. 122

The Importance of Plain Speaking

Use four unknowns: my friend's father's age, my friend's age, and his sister's age, as well as how many years ago my friend's father's age equaled the sum of my friend's and his sister's ages at that time.

Solution: p. 123

16th Week

The Fly

Disregard the fly's complicated flight path; instead, calculate the duration of its flight.

Solution: p. 125

Astonishing Trick

Is it possible to reach twenty with the first seven "taps"? And what about the eighth tap? And the ninth?

Solution: p. 125

17th Week

Impressive Card Trick

Try to see what happens, in its full generality, if the cards are laid out three times.

Solution: p. 129

18th Week

How Did I Know That?

Any three-digit number, with digits a, b, and c, can be written in the form $100a + 10b + c$.

Solution: p. 130

Knight on the Chess Board

With each move, the knight lands on a square of the opposite color.

Solution: p. 131

20th Week

Two Puzzles in Two Languages

Once the numbers to be carried are determined, we get a system of equations for the unknown letters.

Solution: p. 135

Shopping for Chocolate

If one of the young men bought x chocolate bars and his girlfriend bought y, then $x^2 + y^2 = 65$.

Solution: p. 137

21st Week

Instant Addition

Carry out the addition with a and b replacing the two initial numbers.

Solution: p. 138

Still More Jealous Husbands

This "river-crossing problem" is unsolvable. The proof of this is not trivial.

Solution: p. 140

Soccer

There are many ways in which the matches might have been played. Consider how the teams obtained the scores.

Solution: p. 141

22nd Week

Prime Number

Every prime number greater than 3, when divided by 6, leaves the remainder 1 or 5. The puzzle's conclusion follows from this.

Solution: p. 142

Cycle Tour IV

If they agree to two different limits, the player with the greater limit always wins.

Solution: p. 143

Green Cross

We assume that the other applicants are also intelligent young men. Accordingly, exclude all those situations in which a decision would have been obvious.

Solution: p. 144

23rd Week

Money for Ice Cream

Verify that if the last two digits of a square number are the same, then they are either 00 or 44.

Solution: p. 145

The Envious Cousin

Prove that if the last four digits of a square number are the same, then they are all 0.

Solution: p. 146

24th Week

A Ship Sails By

Warning! The solution of the puzzle differs depending on whether the ship is faster or slower than Tommy.

Solution: p. 147

25th Week

Lesson in Economy

Use the relationship between the arithmetic and geometric means (see the solution of "A Journey by Air" (1st Week, p. 90)).

Solution: p. 151

Auto Racing

Compare the time it must have taken the car to reach Northy from Cambridge with the time it used to reach the summit of the mountain.

Solution: p. 153

26th Week

Uncle Pops a Quiz

Use the table of prime factorizations given in Section 5 of Appendix, p. 226.

Solution: p. 154

27th Week

Colored Dice

First, count how many ways the dice can be painted; then count the ways the numbers can be written on the faces of the dice. (Or the other way around!)

Solution: p. 156

28th Week

Randolini's Card Trick

Take a deck of cards and lay out the cards on the table clockwise in a large circle. When the cards have all been laid out, it's no longer possible to tell which was the first card and which was the last; which was the second card, and so on. Which card was just before another, however, is preserved. We say that the *cyclical order* of the cards is retained.

Start with the observation that when we cut the deck, placing the cards cut from the top to the bottom of the deck, but otherwise not changing their order, this does not change the cyclical order.

Solution: p. 158

30th Week

Carton Puzzles

Next to each letter, write the number of different ways one can reach it from the letter C.

Solution: p. 162

Toy Soldiers—Halt!

Setting up in even rows reveals certain divisibilities and remainders, leading to a contradiction.

Solution: p. 165

31st Week

Party Question

Use induction on the total number of friendships.

Solution: p. 167

First and Last Day of the Year

Take into account that whether in one year, two dates fall on the same day of the week depends on the number of days in the year.

Solution: p. 169

32nd Week

Bridge Party

List all possible seating arrangements that satisfy the conditions.

Solution: p. 169

Beside/Below Redux

Observe the order in which the cards were laid out. In the first round, for example, the odd cards were laid out; in the second round, cards with numbers divisible by 4; and so on.

Solution: p. 170

Weight Trick III

Divide the boxes into three groups of four, and apply what we learned from the solution of "Weight Trick I" (18th Week, p. 131).

Solution: p. 171

33rd Week

The Spy Who Came around the Corner

Next to each intersection, write the number of ways by which it can be reached from the designated gate. This yields a part of Pascal's triangle (see the solution of "Carton Puzzles" (30th Week, p. 162)).

Solution: p. 172

34th Week

Battle of Cards

Let w, x, y, and z be the money the four players had at the start of the game. Find equations for these.

Solution: p. 175

Remains of a Long Division

First, determine the divisor.

Solution: p. 177

35th Week

Rook Walk

The reader should revisit the solution of "Knight on the Chess Board" (18th Week, p. 131).

Solution: p. 180

Married Couples

This puzzle is similar to "Shopping for Chocolate" (20th Week, p. 29).

Solution: p. 180

36th Week

Legal Problem from Antiquity

Take into account how many dishes Titus received, and from whom.

Solution: p. 181

Square Numbers

Let a be an even number between two consecutive odd numbers; find the relationship among the three squares.

Solution: p. 182

38th Week

Headscratcher with Caps

Look for the solution, assuming that all three think logically, and that they know this about each other.

Solution: p. 185

Schoolgirls

To get an overview of the situation, draw a 10×10 table. Label the rows with the initials of the names of the girls, and the columns with the grades, I–X. Then cross off the square at the "intersection" of a column and a row if the girl cannot be in that grade.

Solution: p. 185

39th Week

Tennis Tournament

Prove that if we do the report with the winner of the tournament, then we achieve our goal.

Solution: p. 187

Hurdle Race

Start from the facts that, in both contests, there were exactly five possible ways of distributing the gift certificates, and the number of gift certificates was divisible by five.

Solution: p. 188

40th Week

Vacation Days

Use the idea of the solution of "The Metal Tube" (4th Week, p. 95).

Solution: p. 191

41st Week

Divine Injunction

Assume that there are n disks on the first needle, and prove by induction on n (see Section 2 of the Appendix, p. 222) that $2^n - 1$ is the fewest number of steps in which the disks can be transferred from that needle to the second one. By the induction step, determine the moment at which the originally lowermost disk is transferred from the first needle to the second.

Solution: p. 193

Mathematicians in Conversation

The following observation can prove useful in solving this puzzle: let $a_1, \ldots, a_n$ and b be integers, satisfying

$$a_1 + \cdots + a_n = b,$$

and let us assume that with one exception the integer d divides all a_i and b. Then d divides all a_i.

The proof of this observation is simple. Assume that a_1 is the exception, of which it is not yet known whether it is divisible by d. The other a_i and b are divisible by d, so $a_2 = dk_2$, $a_3 = dk_3$, …, $a_n = dk_n$, $b = dk$, for some integers $k_2, \ldots, k_n$, and k.

With this notation, we have

$$a_1 = d\big(k - k_2 - \cdots - k_n\big).$$

Thus, a_1 is, in fact, divisible by d.

The most important relevant observations about number systems are found in the solution of "Curious Addition" (9th Week, p. 106).

Solution: p. 195

42nd Week

Gifts

Write $80 as all possible sums that fulfill the stated conditions.

Solution: p. 196

The Fickle Sultan

To solve this problem, find positive integers that have an odd number of divisors.

Solution: p. 198

43rd Week

Military Band—Blow Your Own Horn!

Proceed from the fact that each officer has two brothers-in-law; therefore, they can be arranged in a cyclical order so that each finds himself between his two brothers-in-law.

Solution: p. 198

At the Round Table

Calculate the number of seating options there are if

(a) all four couples sit next to each other;

(b) three couples sit next to each other, and the fourth couple is separated;

(c) two couples sit next to each other, and the other two couples are separated;

(d) one couple sits next to each other, and the other three are separated.

Then subtract the sum of these possibilities from the total number of seating options.

Solution: p. 199

44th Week

Tommy's Surprise

We prepare for the solution with the following observation:

Let $a \neq 1$ and $b \neq 5$ be single-digit numbers with the property that the product $a \cdot b$ ends with the digit b. Then $a = 6$ and b is one of 2, 4, 6, or 8.

Indeed, if b is the last digit of $a \cdot b$, then

$$ab = 10x + b,$$

where x is a positive integer. We rewrite this as

$$(a - 1)b = 10x.$$

The right-hand side of this equation is divisible by 5, and thus this is true also of the left-hand side. If a product is divisible by 5, then at least one of its factors is also divisible by 5, since 5 is a prime number. The number $b \neq 5$, however, is not divisible by 5. Therefore, 5 is a divisor of $a - 1$, and so, $a = 6$. Substituting this into the last displayed equation, we get $5b = 10x$, that is, $b = 2x$. Thus, b is an even number.

Utilizing this observation, first determine the value of T in the multiplication.

Solution: p. 203

45th Week

A Number in Mind

Not all digits can appear as the last digit of a square number. Based on this, we can determine the last digit of Alfred's number.

The following observation also proves useful: The square of an odd integer divided by 4 leaves the remainder 1. (The square of an even integer, on the other hand, is divisible by 4.)

Indeed, an odd integer is of the form $2n + 1$ for some integer n, and its square is

$$4n^2 + 4n + 1 = 4(n^2 + n) + 1,$$

so divided by 4 leaves the remainder 1. (An even number is of the form $2n$, where n is an integer, so the square, $4n^2$, is obviously divisible by 4.)

Solution: p. 204

The Hungry Sales Woman

We should argue that all three baskets were empty.

Solution: p. 206

46th Week

Return of the Sultan

Search for those positive integers, not greater than 100, with the property that the sum of their divisors is odd. It's easy to reason that the sum of the divisors can be odd only if the number of odd divisors is odd.

To solve this puzzle, it may be useful to reread the solution of "The Fickle Sultan" (42nd Week, p. 198).

Solution: p. 208

47th Week

An Interesting Game

Make use of the symmetry of the table.

Solution: p. 209

49th Week

Footrace

Consider those numbers smaller than 144 that when divided by 13 leave the remainder 1, and that can be written as the product of two numbers, both smaller than 13.

Solution: p. 213

No Tricks

Write out all possible decompositions of 1,296 into the product of three numbers, and calculate the sums of the factors. The solution of the puzzle follows easily.

Solution: p. 213

51st Week

Book Lovers

The identities given in Section 3 of the Appendix, p. 223:

$$a^n - b^n = (a-b)\big(a^{n-1} + a^{n-2}b + a^{n-3}b^2 + \cdots + a^2b^{n-3} + ab^{n-2} + b^{n-1}\big)$$

(n is a positive integer) and

$$a^n + b^n = (a+b)\big(a^{n-1} - a^{n-2}b + a^{n-3}b^2 - \cdots + a^2b^{n-3} - ab^{n-2} + b^{n-1}\big)$$

(n is an odd positive integer), will prove useful.

Solution: p. 216

52nd Week

Weighing Once Wins!

Note that nothing in the formulation of the puzzle precludes opening the crates!

Solution: p. 218

Solutions

1st Week

A Journey by Ship

Let's say my ship left New York at noon on the 10th and arrived in Lisbon at noon on the 18th. Then, the first ship I count left Lisbon at noon on the 2nd, and the last ship is leaving just as my ship arrives at noon on the 18th. We are, at this time, 16 days (or 32 half days) into our journey. Since we start our count at the time of departure and conclude it at the time of arrival, I counted 33 ships traveling in the opposite direction.

Two Games of Chess

My sister sat my two friends at separate tables, which we'll call Table One and Table Two. My sister played black against Tom at Table One (and hence went second) and played white against Paul at Table Two.

First, she went to Table One and invited Tom, who played white, to start. Tom made a move. My sister went to Table Two, where she played white, and made exactly the same opening move. Then she waited until Paul made his counter move, went back to Table One, and made exactly the same move on the first chess board.

She continued playing the same way—so her opponents sitting at the two tables in effect played each other, and my sister was an intermediary, conveying their moves back and forth between tables.

Thus, she achieved a better record than my two losses. Either both games ended in a draw, or my sister won one game and lost the other.

A Journey by Air

Our conclusion is logical but incorrect.

The wind speed was 30 mph, and thus we flew from Boston to New York City at a speed of $150 + 30 = 180$ mph. On the return flight, our speed was $150 - 30 = 120$ mph. We calculate the duration of the flight by dividing the distance by the speed. Thus, the return flight took (in hours):

$$\frac{200}{180} + \frac{200}{120} = 2 + \frac{7}{9} = 2.777\ldots.$$

If the wind speed is zero, the flight would take

$$\frac{400}{150} = 2 + \frac{2}{3} = 2.666\ldots$$

hours. Thus, we lost about $2.777 - 2.666 = 0.111$ hours, that is, 6.66 minutes because of the wind.

Note that given *any* wind, the duration of the flight increases. If the wind speed is x mph, where x is any positive number, then the duration of the flight in hours is

$$\frac{200}{150 + x} + \frac{200}{150 - x} = 200\left(\frac{1}{150 + x} + \frac{1}{150 - x}\right).$$

Our claim holds because the inequality

$$\frac{1}{150 + x} + \frac{1}{150 - x} \geq \frac{2}{150} \tag{1}$$

is true. This inequality becomes an equality only for $x = 0$. Note that the right-hand side is what we get when $x = 0$, that is, when no wind is blowing.

To verify the inequality (1), we can assume that $x < 150$, since otherwise the aircraft could not even have taken off from New York City. Multiplying both sides of the inequality (1) by $150(150 + x)(150 - x)$, we get that

$$150^2 \geq (150 + x)(150 - x). \tag{2}$$

By multiplying out the right-hand side,

$$150^2 \geq 150^2 - x^2,$$

we can see that the inequality (1) holds.

* * *

Inequality (2) is a special case of the familiar inequality between arithmetic and geometric means. This inequality states that if a and b are positive numbers, then

$$\frac{a+b}{2} \geq \sqrt{ab} \tag{3}$$

and equality exists only for $a = b$. If we set $a = 150 + x$ and $b = 150 - x$, we obtain, in fact, inequality (2).

Now let us prove the inequality between the arithmetic and the geometric means. The square of any real number is not negative, and therefore, it is true that

$$(\sqrt{a} - \sqrt{b})^2 \geq 0,$$

with equality holding only for $a = b$. Expanding the left-hand side gives

$$a + b - 2\sqrt{ab} \geq 0,$$

from which inequality (3) follows by rearranging the terms and then dividing both sides by 2.

2nd Week

The Math Whiz

The method described in the puzzle can always be used, as can be seen with the help of a little computation.

We can write a fraction whose numerator is one less than its denominator in the form

$$\frac{a-1}{a} = 1 - \frac{1}{a},$$

where a is an integer ($a \neq 0$, since a fraction cannot have 0 as the denominator).

Compute the difference of two such fractions, for integers a and b (both nonzero):

$$\frac{a-1}{a} - \frac{b-1}{b} = \left(1 - \frac{1}{a}\right) - \left(1 - \frac{1}{b}\right) = \frac{1}{b} - \frac{1}{a} = \frac{a-b}{ba}.$$

The general rule, then, is as follows: subtract the first denominator from the second denominator, to obtain the new numerator, and multiply the denominators to obtain the new denominator. Note that you can also get the new numerator by subtracting the original first numerator from the original second numerator, since $a - b = (a-1) - (b-1)$.

The Will

The Qadi advises them each to mount the camel of the other, in order to reach the oasis on the brother's camel as quickly as possible. Thus, the *camel* of the brother who reaches the oasis first will arrive there later, winning this brother the inheritance of his father's treasures.

The Clock

I started up the clock to determine how long I was away. If I subtract the time I spent with my friend from the time I was away, I obtain the time it took me to go there and back. I add half of this time to the time when I left my friend's house (which I noted on my friend's clock as I left). On my return home, I adjust my wall clock to this time.

Let's now consider this little calculation algebraically as well. Let A be the time my wall clock showed at my departure. At my friend's house, I noted the exact time of my arrival and my departure; let h and k denote these. When I returned home, my wall clock showed the time B. So $B - A$ is the time I was away. Of this, I spent the time $k - h$ at my friend's house. Let t be the time I need to walk to my friend's house (and the time it takes to return home). Then

$$2t = (B - A) - (k - h).$$

Therefore, $k + t$ is the time of my return, and I set my clock accordingly. (We assume that I have traveled there and back at the same speed.)

3rd Week

Nicholas's House

We must begin at one of the two bottom corners of the house and end at the other. This is because these two points (labeled 1 and 2 below) are the

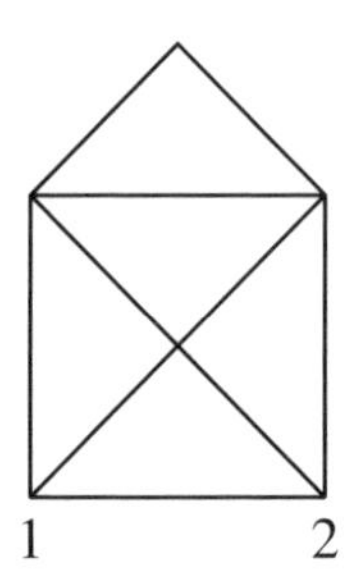

meeting points of an odd number of lines (three in each case), while every other point is the meeting point of an even number of lines.

If Point 1, for example, is neither the starting nor the ending point, we would arrive at Point 1 exactly as often as we would leave it. Therefore, each passage through Point 1 would "pair" the lines meeting at Point 1 and so the number of lines meeting at Point 1 would be even. This isn't the case, however; thus, Point 1 (and Point 2) can be only the starting or ending point.

The rule: start at Point 1 or Point 2.

By following the rule, we'll never get stuck drawing the lines of the house.

An Unusual Year

We assign the number 1 to the 13th of January, and the numbers 2, 3, ..., 7 to the following six days. If, for example, the 13th of January was a Wednesday (to which we assign the number 1), then we assign the number 2 to Thursday, ..., the number 7 to the following Tuesday. Then we continue to assign the numbers 1, ..., 7 cyclically, to all days of the year, ending with the 12th of January.

How can we easily figure out which number is assigned to the 13th of February? If January had 28 days, we would have assigned the 13th of February the same number as the 13th of January, namely, the number 1. But since January has 31 days, 3 more than 28, therefore, the number assigned to the 13th of February is $1 + 3 = 4$. In general, if a month has x days, and if x divided by 7 leaves the remainder r, then the numbers assigned to the days of the next month are increased by r.

From this and the fact that the year in question was a leap year, we can determine the numbers assigned to the 13th of each month:

13th of month	number assigned
January	1
February	4
March	5
April	1
May	3
June	6
July	1
August	4
September	7
October	2
November	5
December	7

We see there is only *one* number that occurs three times, namely, the number 1. Consequently, the number 1 was assigned to Friday, because in Kate's birth year, the 13th fell on a Friday three times. And since April 13th was a Friday, April 1st was a Sunday. Thus, Kate was a Sunday's child.

* * *

Now let's see how the numbers assigned to the 13th of the month change when the year is not a leap year. Obviously, January and February remain unaffected, while the other numbers are each reduced by 1. Thus, we get

$$1,\ 4,\ 4,\ 7,\ 2,\ 5,\ 7,\ 3,\ 6,\ 1,\ 4,\ 6.$$

We now observe another interesting point. In both the leap year and in the "regular" year (that is, the non-leap year), all the numbers appear in the list of days on which the 13th falls. From this—among other things—follows the interesting fact that in every year, the 13th of the month falls at least once on a Friday.

Competitors, or *The Raise*

The second option yields a higher income, so the third candidate—who chose it—was hired.

To see that the third candidate was right, look at the following table of income from the two options:

	Option 1	Option 2
1st month	2,000	1,000 + 1,050 = 2,050
2nd month	2,150	1,100 + 1,150 = 2,250
3rd month	2,300	1,200 + 1,250 = 2,450

Option 2 is thus more lucrative.

Let us calculate this algebraically. Let n be the base salary. Let a be the monthly salary increase in Option 1; let b be the bimonthly salary increase in Option 2. In Option 1, the monthly salaries are

$$n, \quad n+a, \quad n+2a, \quad n+3a, \ldots.$$

In Option 2, the bimonthly checks add up to the monthly incomes:

$$n+b, \quad n+5b, \quad n+9b, \quad n+13b, \ldots.$$

So in Option 1, the monthly increase in income is a, while in Option 2 it is $4b$. Therefore, if a is less than $4b$, then Option 2 is the more advantageous. If $a = 4b$, the salary increases are the same, but Option 2 provides b dollars more per month.

Thus, we see that a bimonthly increase of $50 is better than a monthly increase of $150.

4th Week

The Metal Tube

Assume that the length of the entire tube is 85 in. so that a 1 in. piece weighs 1 lb. To cut the tube so that both parts weigh an integer multiple of 1 lb, it's necessary to cut at the places where a measuring tape gives integer values. Since the 0 in. and 85 in. marks are at the opposite ends of the tube, it can be cut at the marks of 1, 2, ..., 84. Accordingly, the tube can be cut as desired in 84 ways. (If the tube is symmetric, only 42 of these will produce different results.)

The Ancient Greek Copper Lion

Let x be the time (in hours) required to fill the basin if water flows from all four orifices.

Then in the x hours, the right eye fills $x/24$-th of the basin (using 12-hour days); the left eye, $x/36$-th; the right knee, $x/48$-th; and the lion's mouth, $x/6$-th. Since together they fill the basin in x hours, the sum of the four quotients must equal 1:

$$\frac{x}{24} + \frac{x}{36} + \frac{x}{48} + \frac{x}{6} = 1.$$

Therefore,

$$x \cdot \left(\frac{1}{24} + \frac{1}{36} + \frac{1}{48} + \frac{1}{6}\right) = x \cdot \frac{6+4+3+24}{144} = x \cdot \frac{37}{144} = 1.$$

It follows that

$$x = \frac{144}{37} = 3 + \frac{33}{37}.$$

And thus the basin is filled in about 3 hours and 54 minutes. (Almost 5 hours in our reckoning.)

The Arab

The Quadi suggested that he add a horse of his own to the herd. Then he would give half of the herd of 20 horses, that is, ten horses, to the eldest son; the middle son would receive one fourth, that is, five horses; and the youngest son would get one fifth, or four horses. The horse that would be left over—his own—he would take back home with him. The scheme of the Qadi was ingenious, but it clearly reflected more the spirit than the letter of the will.

5th Week

On the Streetcar

There is but one logical explanation: the passenger gave the conductor two 50-fillér coins.

Hunting Hares

We stipulated that the greyhound makes 8 jumps for every 11 hare-jumps; also, the length of one greyhound-jump equals the length of $9/5$ hare-jumps. Therefore, for every 11 hare-jumps, the greyhound makes 8, of length $72/5 = 14.4$ hare-jumps. So for every 11 hare-jumps, the greyhound gains on the hare the length of 3.4 hare-jumps. Thus, he makes up his deficit of $10 \cdot 3.4 = 34$ hare-jumps in $10 \cdot 11 = 110$ hare-jumps, which equals in length 80 greyhound-jumps.

* * *

We can easily solve the problem with an equation with one unknown. Let x be the number of jumps the hare makes as the greyhound catches up with him. While the hare makes x jumps, the greyhound makes $8x/11$ jumps, which equals

$$\frac{9}{5} \cdot \frac{8x}{11} = \frac{72x}{55}$$

hare-jumps in length. Thus,

$$34 + x = \frac{72x}{55}.$$

It follows that $x = 110$, and so the number of jumps the greyhound makes is

$$8 \cdot \frac{110}{11} = 80.$$

The hare, then, makes 110 jumps before the greyhound, in 80 jumps, catches him.

Telephone Cable—A Long Line?

It is well known that the circumference of a circle of radius r is $2\pi r$, or approximately $6.28r$ (for our present purpose, this rounded figure will suffice).

Let r be the radius (in feet) of the circle formed by the telephone cable. The engineer wants to reposition the line to a circle of radius $r+2$. The additional cable would then be of length

$$6.28(r+2) - 6.28r = 2 \cdot 6.28 = 12.56.$$

The line would thus be 12.56 feet longer, at a cost of $12.56 \cdot 20 = \$251.20$. (Calculated with a more precise value of π, we get \$251.33. Even that expense can be borne!) Interestingly, this number (12.56) is the same whatever the value of r is!

6th Week

■■■■■■□□

Dinner of Dumplings

Let's work our way backwards. Each tourist left two-thirds of the dumplings he found on the plate. So, if the third tourist left 8 dumplings, he must have found $8 \cdot \frac{3}{2} = 12$, therefore, the second tourist left 12. Thus, the second tourist must have found 18 when he woke, which means that the first tourist left 18. And thus, the first tourist must have found 27 dumplings when he woke, which must be the number the host brought out. In other words, the host served

$$8 \cdot \frac{3}{2} \cdot \frac{3}{2} \cdot \frac{3}{2} = 27$$

dumplings.

*   *   *

The problem can also be solved by working our way forward with an equation. Let x be the number of dumplings on the plate served by the host. Then the first tourist ate $x/3$ dumplings, so that

$$x - \frac{x}{3} = \frac{2x}{3}$$

dumplings remained on the plate. Of these, the second tourist ate a third, that is, $\frac{2x}{9}$ dumplings. So,

$$\frac{2x}{3} - \frac{2x}{9} = \frac{4x}{9}$$

dumplings were left for the third tourist, and of these, he ate a third, or $\frac{4x}{27}$ dumplings. When the third tourist was finished,

$$\frac{4x}{9} - \frac{4x}{27} = \frac{8x}{27}$$

dumplings remained. We know that 8 dumplings remained, so

$$\frac{8x}{27} = 8,$$

that is,

$$x = 27.$$

Thus, there were 27 dumplings on the plate to start with.

* * *

This solution can also be illustrated with a table:

	Number of dumplings before meal	Number of dumplings eaten	Number of dumplings after meal
Tourist 1	x	$\frac{x}{3}$	$x - \frac{x}{3} = \frac{2x}{3}$
Tourist 2	$\frac{2x}{3}$	$\frac{2x}{3} \cdot \frac{1}{3} = \frac{2x}{9}$	$\frac{2x}{3} - \frac{2x}{9} = \frac{4x}{9}$
Tourist 3	$\frac{4x}{9}$	$\frac{4x}{9} \cdot \frac{1}{3} = \frac{4x}{27}$	$\frac{4x}{9} - \frac{4x}{27} = \frac{8x}{27}$

Since 8 dumplings were left, therefore, $\frac{8x}{27} = 8$, and so $x = 27$. Thus, there were originally 27 dumplings on the plate.

The Paradox of Protagoras

On Protagoras's side, we can argue that if he wins the trial, then by the decision of the court, he should receive the money due him. If he loses the trial, then his student wins his first trial, and, by their original contract, must pay his teacher's fee.

The student, on the other hand, argues as follows: If I win, then by the decision of the court, I don't have to pay. But if I lose, then by the terms of our contract, I don't have to pay.

The ambiguity of the conclusion arises, of course, from the fact that both teacher and student choose between the decision of the court and the contractual agreement whichever is in their favor. If we stipulate that in either case it is the decision of the court that will be definitive, we can give an unambiguous answer.

* * *

In epistemology[6], relativity is central. Thus, for example, Protagoras (ca. 480–410 BCE) wrote, "Two contradictory statements can be made about everything." That is, in one situation, one statement is true, and in another situation, the other statement is true. From this, he finally draws the conclusion that there is no such thing as an objective situation. This in turn leads to the "*homo mensura* statement" of Protagoras: "Man is the measure of all things ...."[7] Accordingly, from the human viewpoint, not even existence is regarded as objective, but rather as merely a subjective phenomenon.

Say How Many Flags Are Flying

(a) Let's first see how many tricolor flags can be made with four different colors. There are four possible colors for the left stripe. Once the left stripe has been colored, only three colors remain possible for the middle stripe; the color chosen for the left stripe is no longer available. Since we may use every color with every other color, there are $4 \cdot 3 = 12$ different possibilities for the coloring of the left and middle stripes. Once the left and the middle stripes have been colored, only two colors remain for the right stripe. Altogether, then, we can make $12 \cdot 2 = 24$ flags.

(b) If the outer stripes can be the same color, however, then there are four possibilities for the coloring of these two stripes; both stripes red, or both blue, or both green, or both yellow. In all these cases, the middle stripe can be colored in three different ways, that is, altogether $4 \cdot 3 = 12$ flags can be made, of which the two outer stripes are the same color.

* * *

These two problems can both be solved with short arguments. If we knew the fundamental principles of the theory of combinations, however, we could have worked out these problems even more easily. Since similar problems often come up, we take a look at some basic principles.

[6]Theory of knowledge. A branch of philosophy concerned with the nature and scope (limitations) of knowledge.

[7]*Omnium rerum mensura homo.*

> In how many different ways can we arrange six books on a shelf?

Answer: There are 720 different arrangements.

How can we figure this out quickly?

> In general, in how many ways can we arrange n objects in a sequence?

Answer: The number of possible arrangements is $1 \cdot 2 \cdot \cdots \cdot n$. We denote this number by $n!$, and call it n *factorial*; we obtain it by multiplying all the integers from 1 to n.

For example, 3 factorial is $3! = 1 \cdot 2 \cdot 3 = 6$, and 5 factorial is

$$5! = 1 \cdot 2 \cdot 3 \cdot 4 \cdot 5 = 120.$$

We call an arrangement of n different objects in a sequence a *permutation* (of these objects). In formal mathematical language, the answer to the general question is that *n different objects have $n!$ permutations.*

We can prove this by induction. (The principle of induction is explained in Section 2 of the Appendix, p. 222.)

Start with the case that $n = 1$: one object can be arranged in only one way. Since $1! = 1$, the statement is true for $n = 1$.

We assume, now, that the statement is true for $n - 1$. We separate the permutations of n elements into classes: We place two permutations into the same class if they begin with the same element. Since we are dealing with n elements, we get n different classes. In each class, the permutations have the same first element, and therefore, the number of different permutations found in each class equals the number of permutations of the remaining $n - 1$ elements. On the basis of the induction hypothesis, there are $(n - 1)!$ elements in every class. Altogether there are n classes, and thus $(n - 1)! \cdot n = n!$ is the number of permutations of n objects.

Another type of question that often arises is the following:

> How many k-color flags can we make with n colors (we assume that the flags consist of colored stripes, and any two stripes in a flag must be of different colors)?

Answer: The number of possibilities is given by the product

$$n(n - 1)(n - 2) \cdots (n - k + 1).$$

With the factorial notation, we can express this more simply:

$$\frac{n!}{(n - k)!}.$$

We can derive this formula as we did in case $n = 4$, $k = 3$ in the solution of the puzzle.

If we select k elements from n elements such that we also take into account the sequence of the selected elements, then we speak of *arrangements* of k elements from a set of size n. Another formulation of the previous statement is that the number of arrangements of k elements, taken from a set of n elements, is

$$\frac{n!}{(n-k)!}.$$

Often, though, the sequence of the selected elements is irrelevant.

> The three winners of a prize of $6,000 offered by a newspaper are selected by a drawing from among the 15 people who submitted the correct solution to a puzzle. How many different groups of three winners can be selected?

Answer: There are 455 possible results of the drawing.

In this case, it is irrelevant in what order the three names are drawn. All that matters is which three names have been drawn. The selection of some elements from among a greater number of elements, irrespective of their order, is called a *combination*. With the help of the observations above, the number of combinations of k elements from a collection of n elements (in other words, the number of combinations of k from n) can easily be determined. Let x be this number. If we order the selected elements in all possible ways, that is, if we permute them, then altogether we have $k! \cdot x$ possibilities. But that total is exactly the same as the number of arrangements of k elements from a set of size n, which is

$$\frac{n!}{(n-k)!}.$$

Thus, we have that

$$k! \cdot x = \frac{n!}{(n-k)!},$$

and from this it follows that

$$x = \frac{n!}{(n-k)!\,k!}.$$

This expression is so important that a separate symbol has been introduced to represent it: $\binom{n}{k}$. We call this expression "n choose k." Accordingly,

$$\binom{n}{k} = \frac{n!}{(n-k)!k!}.$$

7th Week

■■■■■■■□□

Homework

Let the letters a, b, c, d, and e represent digits. Then, the first multiplication problem is

$$\overline{abcde7} \cdot 5 = \overline{7abcde}$$

(By overscoring, we indicate that the digits represent a number written in the decimal system.) Since $5 \cdot 7 = 35$, we have $e = 5$ with 3 to carry. It follows that $5e + 3 = 5 \cdot 5 + 3 = 28$, and hence $d = 8$, with 2 to carry. If we continue in this manner, we get $c = 2$, $b = 4$, and $a = 1$. Therefore, the first multiplication problem is

$$142{,}857 \cdot 5 = 714{,}285.$$

We could have solved this problem by another method, which we'll use with the second multiplication. The second multiplication problem is

$$\overline{1abcde} \cdot 3 = \overline{abcde1}.$$

Let x be the five-digit number $\overline{abcde}$. Then

$$(100{,}000 + x) \cdot 3 = 10x + 1,$$

and from this it follows that $7x = 299{,}999$; therefore, $x = 42{,}857$. Hence, the product is

$$142{,}857 \cdot 3 = 428{,}571.$$

The Superhotel

The error is that we have counted the second guest also as the seventh guest.

Jealous Husbands

The solution is straightforward.

1. One couple crosses the river.

2. The husband returns.

3. The men cross.

4. The second husband returns.

5. The second couple crosses.

The puzzle has a second solution. (Find it!)

The story fails to report, however, whether the wives, left alone, remained faithful to their husbands with some man cast by chance on the shore.

8th Week

Another Superhotel

The question is formulated incorrectly.

If we subtract 20 dollars from 270 dollars: $270 - 20 = 250$, we get the amount the proprietor received for the apartment.

* * *

Put another way, each tourist paid

$$\frac{250}{3} = 80 + \frac{10}{3}$$

dollars for a room.

Each tourist received ten dollars back. That makes

$$90 + \frac{10}{3}$$

dollars each.

The bellhop put $20/3$ dollars from each of the tourists into his pocket, which altogether makes

$$3 \cdot \frac{20}{3} = 20$$

dollars.

If we add up these amounts,

$$3 \cdot \left(90 + \frac{10}{3}\right) + 3 \cdot \frac{20}{3},$$

we get exactly 300 dollars.

Three-Letter Words

In the first addition, $N + R = N$, that is, $R = 0$, and there is nothing to carry. $A + I = R = 0$ is impossible, since if it holds, then $A = I = 0$, contradicting that $A \neq I$. Thus, we have

$$A + I = 10. \tag{1}$$

We have to carry the 1, which gives

$$C + S + 1 = U. \tag{2}$$

In the second addition in the middle column, N is below R and A, that is, under 0 and A. Since A and N are different digits, only

$$A + 1 = N \tag{3}$$

is possible. Therefore, 1 has to be carried from the previous column, that is,

$$2N = E + 10. \tag{4}$$

By (3), A is at most 8, and thus $R + A + 1$ leaves nothing to be carried (in this case $A = 9$, $N = 0$, which isn't possible because $N \neq R$). This means that

$$U + C = O. \tag{5}$$

From (3) and (4) it follows that $2N = 2(A + 1) = E + 10$, and thus

$$A = \frac{E}{2} + 4. \tag{6}$$

We know that $A \neq 9$, and since $E \neq 0$ (since $R = 0$), A must be at least 5. However, $A = 5$ is not possible, since then $I = 5$ according to (1); $A = 6$ isn't possible, since otherwise $I = 4$ according to (1), and $E = 4$ by (6); $A = 8$ is also impossible, since otherwise $E = 8$ by (6).

Thus, we have $A = 7$. Substitutions in (1), (3), and (6) yield $I = 3$, $N = 8$, and $E = 6$. We now know the following about the additions:

$$\begin{array}{rccc} & C & 7 & 8 \\ + & S & 3 & 0 \\ \hline & U & 0 & 8 \\ + & C & 7 & 8 \\ \hline & O & 8 & 6 \end{array}$$

From (2) and (5), we obtain that

$$2C + S + 1 = O. \tag{7}$$

If C were greater than 3, then by (7) (since S is greater than 0), the value of O would necessarily be at least 10, which is impossible; $C = 3$ is impossible because $I = 3$.

We assume that $C = 2$. Then, because of (7),

$$O = S + 5.$$

Accordingly, O is at least 6. But since 6, 7, and 8 are already in use (as E, A, and N, respectively), only $O = 9$ would be possible; in this case, though, because of (5), $U = 7$ would hold and would thus be the same as A, a contradiction.

Accordingly, $C = 1$. Because of (7), it follows that

$$O = S + 3. \tag{8}$$

By (5), and taking into account which digits have yet to be identified, only $S = 2$ and $O = 5$ fulfill condition (8). Because of (5), then, we have $U = 4$. Therefore, only one solution is possible:

$$\begin{array}{r} 1\ 7\ 8 \\ +\ 2\ 3\ 0 \\ \hline 4\ 0\ 8 \\ +\ 1\ 7\ 8 \\ \hline 5\ 8\ 6 \end{array}$$

In the Kindergarten

Separate the children into four groups:

1. blue eyes and brown hair,
2. blue eyes and black hair,
3. green eyes and brown hair,
4. green eyes and black hair.

We now select one child—Tommy—and assume that Tommy belongs to Group 1. If there is a child who belongs to Group 4, then Tommy and that child form a pair we seek.

However, what if there is no child with green eyes and black hair? In that case, there is certainly a child who belongs in Group 3, since otherwise there would be no one in the kindergarten with green eyes, contrary to what we assumed, namely, that in this kindergarten both eye colors and both hair colors are present. But there is also a child who belongs in

Group 2; otherwise, there would be no black-haired children in the kindergarten.

We will find the pair we are looking for by selecting a child from Group 2 and a child from Group 3. We could argue similarly had we initially chosen a child from a different group.

9th Week

How Old Is Susie?

Let Susie's mother be x years old, her father y years old, and Susie z years old. Then for the prime numbers x and y, and for the integer z, the following two equations hold:

$$x + y = 100,$$
$$x = \frac{y + z}{2}.$$

From the latter it follows that $z = 2x - y$. We also assume that x and y (the ages of the parents) are both greater than z (the age of the Susie) by at least 12.

The number 100 can be written as the sum of two primes in several ways, and thus there are several candidates for x, y, and z. We eliminate the cases that yield a negative number for z and also cases in which Susie would be older than her parents. That leaves the following two possibilities:

$$x = 41,\ y = 59,\ z = 23,$$
$$x = 47,\ y = 53,\ z = 41.$$

The second fails the condition that the age difference between Susie and her parents is at least 12 years. Thus, Susie's father is 59 years old, her mother is 41, and Susie herself is 23.

Curious Addition

The Professor told his wife he had carried out the addition in base 9, that is, in the number system in which the places represent powers of 9.

* * *

Normally, as we all know, we operate in the base 10 number system, called the *decimal system*. We know, too, what a significant achievement the introduction of the decimal system was at the time. Previously, in many places, other number systems were in use, whose traces we still encounter today. For example, the *dozen* (12) and the *gross* ($12 \cdot 12 = 144$) are remnants of a base 12 system, while the division of an hour into 60 minutes, and the subdivision of a minute into 60 seconds, are remnants of the base 60 (*sexagesimal*) system (that is, the number system whose place values represent powers of 60).

What really is a number system?

Let's say I have bought a gift for \$1,265. What does "1,265" symbolize? The digits have a *place value* and a *nominal value*. The 5 means five ones, the 6 six tens, the 2 two hundreds, and the 1 one thousand:

$$1{,}265 = 1 \cdot 1{,}000 + 2 \cdot 100 + 6 \cdot 10 + 5.$$

The 2, for example, has the nominal value 2 and the place value 100. We shall call \$1,265 a *four-digit* number (often also called a *four-place* number).

Similarly, we have

$$23{,}654 = 2 \cdot 10{,}000 + 3 \cdot 1{,}000 + 6 \cdot 100 + 5 \cdot 10 + 4.$$

To avoid writing so many zeros, we use *exponentiation*, that is, we write 100 as 10^2, 1,000 as 10^3, and 10,000 as 10^4. We can also write 10 as 10^1 (10^n means that 10 is multiplied by itself n-times). For convenience, we write $10^0 = 1$.

By placing a digit in the second, third, fourth, ..., nth place from the right, we express that these digits have 10-times, 10^2-times, 10^3-times, ..., 10^{n-1}-times their nominal value.

For example, 1,265 can be written with exponentiation as

$$1{,}265 = 1 \cdot 10^3 + 2 \cdot 10^2 + 6 \cdot 10^1 + 5,$$

or

$$1{,}265 = 1 \cdot 10^3 + 2 \cdot 10^2 + 6 \cdot 10^1 + 5 \cdot 10^0.$$

In general,

$$\overline{a_n a_{n-1} \dots a_2 a_1 a_0} = a_n \cdot 10^n + a_{n-1} \cdot 10^{n-1} + \cdots + a_2 \cdot 10^2 + a_1 \cdot 10^1 + a_0.$$

(The overscore on the left-hand side of the equation means that the a_i in the corresponding sequence indicates the digits of a number written in the decimal system; this is an $(n+1)$-*digit* number.)

We can write the numbers in an analogous fashion in a number system other than base 10. In base 3, for instance, we can write 75 in the following way:

$$75_{10} = 2 \cdot 3^3 + 2 \cdot 3^2 + 1 \cdot 3^1 + 0 = 2210_3.$$

The small 10 next to the 75 (called a subscript) means that we have an expression in base 10. The same way, writing subscripts 3, 4, ... indicates that the expression is in base 3, base 4, ... In base 3, we write the numbers as powers of 3 and use only the digits 0, 1, and 2. In base 3, there are fewer digits, but the numbers get longer. For this and other reasons of convenience, we'll generally stick to calculations in the decimal system. (Note here that from a practical standpoint, the base 12 (*duodecimal*) system would be preferable to the decimal system, because 12 is divisible by 2, 3, 4, and 6.)

Let's see now how a number in base b is written. We write the number in the form

$$a_n \cdot b^n + a_{n-1} \cdot b^{n-1} + \cdots + a_2 \cdot b^2 + a_1 \cdot b + a_0,$$

where we select each of the digits a_i from among the numbers 0, 1, ..., $b - 1$.

The number 16 in the decimal system is expressed in base 3 as follows:

$$16_{10} = 1 \cdot 3^2 + 2 \cdot 3 + 1 = 121_3.$$

The same number in base 2 (*binary*) is

$$16_{10} = 1 \cdot 2^4 + 0 \cdot 2^3 + 0 \cdot 2^2 + 0 \cdot 2^1 + 0 \cdot 2^0 = 10{,}000_2.$$

The following two questions arise: Is it true that every number can be expressed in a number system in base b, where b is any natural number greater than 1? And is such an expression unique?

For simplicity's sake, we'll confine ourselves here to the case of base 3 (the arguments used are valid for all other bases). Let's write out 181 in base 3.

We divide 181 by 3:

$$181 = 3 \cdot 60 + 1,$$

and the remainder 1 is the last digit of this number in base 3. We proceed in the same manner with the remaining quotients:

$$\begin{aligned} 60 &= 3 \cdot 20 + 0, \\ 20 &= 3 \cdot 6 \;\; + 2, \\ 6 &= 3 \cdot 2 \;\; + 0, \\ 2 &= 3 \cdot 0 \;\; + 2. \end{aligned}$$

Thereby, we get in reverse order the digits in base 3—that is, 181_{10} is $20{,}201_3$. This is, in fact, correct, since

$$181_{10} = 2 \cdot 3^4 + 0 \cdot 3^3 + 2 \cdot 3^2 + 0 \cdot 3^1 + 1 = 20{,}201_3.$$

The correctness of this procedure is readily seen. To see the uniqueness of the expression, let

$$181_{10} = a_4 \cdot 3^4 + a_3 \cdot 3^3 + a_2 \cdot 3^2 + a_1 \cdot 3^1 + a_0.$$

Then the division of 181 by 3 obviously leaves a_0 as the remainder, and that is the last digit. We also see that the quotient is

$$a_4 \cdot 3^3 + a_3 \cdot 3^2 + a_2 \cdot 3 + a_1.$$

Continuing this process, we get a_1 is the next remainder, and so on, and we obtain a_2, a_3, and a_4.

Let's now consider another example: we'll write 38 in the binary (base 2) system. Carrying out the divisions, we obtain

$$\begin{aligned} 38 &= 2 \cdot 19 + 0, \\ 19 &= 2 \cdot 9 + 1, \\ 9 &= 2 \cdot 4 + 1, \\ 4 &= 2 \cdot 2 + 0, \\ 2 &= 2 \cdot 1 + 0, \\ 1 &= 2 \cdot 0 + 1. \end{aligned}$$

And thereby, $38_{10} = 100{,}110_2$.

We see that we've gotten much more than we'd bargained for—not just the unambiguous expression we sought, but also a handy procedure for obtaining such a representation.

Now we're in a position to easily check whether the Professor has calculated correctly on his piece of paper. The sum of the digits in the last column in base 9 is 12_9; we write down the 2, which leaves the 1 to carry. The sum in the next column is 25_9, and with the 1 carried over we obtain 26_9. We write down the 6, and there remains 2 to carry .... In this way, we can check the calculation and confirm that the addition was correct.

* * *

A further question arises in connection with this puzzle: can we deduce that the addition was indeed in base 9? First, let's write out the four numbers in the puzzle as base q numbers:

$$\begin{aligned} 873_q &= 8q^2 + 7q + 3, \\ 381_q &= 3q^2 + 8q + 1, \\ 1184_q &= 1q^3 + 1q^2 + 8q + 4, \\ 3_q &= 3, \\ 2562_q &= 2q^3 + 5q^2 + 6q + 2. \end{aligned}$$

The sum of the right-hand sides of the first four equations equals the right-hand side of the last equation. That is,

$$8q^2 + 7q + 3 + 3q^2 + 8q + 1 + q^3 + q^2 + 8q + 4 + 3 = 2q^3 + 5q^2 + 6q + 2.$$

Rearrangement yields

$$q^3 - 7q^2 - 17q - 9 = 0.$$

From this it follows that the q we are looking for is a root of this third-degree equation. We use the following well-known theorem: if in an n-degree equation with integer coefficients the leading coefficient equals 1, then the integer roots of the equation divide the constant term. In our case, this means we must be dealing with a number system whose base is a divisor of 9. Since $q = 3$ is not a solution of our third-degree equation, $q = 9$ is the only possible base.

We still need to check whether $q = 9$ is a solution of the equation in question. This is indeed the case, since

$$9^3 - 7 \cdot 9^2 - 17 \cdot 9 - 9 = 0.$$

And accordingly, the addition on the paper was, in fact, written in base 9.

* * *

The above procedure utilized a cubic equation; alternatively, we could use a linear equation. The sum of the last column,

$$3 + 1 + 4 + 3 = 11,$$

can be written in a base x as $nx + 2$, that is, $11 = nx + 2$, and from this it follows that $nx = 9$. But x is at least 9, since 8 appears as one of the digits. Thus, $x = 9$.

* * *

We'll encounter the binary and the base 3 systems as we proceed. These number systems will serve us well in solving puzzles that seem to have nothing to do with such matters.

Exercise in Logic

The second and third statements can be simultaneously true. If the second statement is true, then the third is also true. The second and third statements can also be simultaneously false, in the case that the first statement is true.

The first and second statements can both be false, in case Bacon wrote no play of Shakespeare's. It is, however, obvious that these two statements cannot simultaneously be true.

10th Week

■■■■■■■■■■□□

The Local Lurches On

Since the express train traveling at twice the speed of the local catches up with it at Washington, DC, the local has traveled half the distance when the express departs. And when the local has traveled $2/3$ of the distance, the express has traveled $1/3$. The local train has covered

$$\frac{2}{3} - \frac{1}{2} = \frac{1}{6}$$

of the distance from the time the express departed, that is, in the time it has taken the express to travel $1/3$ of the way. After that point, the local's speed is reduced to half of its original speed, meaning the express is traveling at four times the local's speed.

If we take the distance from Richmond to Washington, DC, to be one unit, and we let x be the distance the local covers between the time the mechanical problem arises to the point in time at which the express catches up with it, then

$$\frac{2}{3} + x = \frac{1}{3} + 4x,$$

that is,

$$3x = \frac{1}{3},$$

and thus

$$x = \frac{1}{9}.$$

From the point the mechanical problem arose, the local had only $1/3$ of the distance still to go, and therefore, at the time the express caught up, there remained a distance of

$$\frac{1}{3} - \frac{1}{9} = \frac{2}{9}$$

to go to Washington, DC; by assumption, this distance is

$$27 + \frac{1}{9} = \frac{244}{9} \text{ miles.}$$

The distance from Richmond to Washington, DC, is, then,

$$\frac{244}{9} \cdot \frac{9}{2} = 122 \text{ miles.}$$

Cocoa Island

The solution rests on the fact that, to the question "Are you Bodo or Wizzer?" every native—regardless of whether he is Bodo or Wizzer—can give only the answer "Bodo."

For if he is Bodo, then he tells the truth, and answers "Bodo." If he is Wizzer, however, he lies, and therefore, his answer is also "Bodo."

So Abl has said that he is "Bodo," which means that Bislu has lied and is accordingly a Wizzer. Cacil, on the other hand, has told the truth and is thus a Bodo.

Note that we cannot decide whether Abl is Bodo or Wizzer.

The Little Troublemaker

If similar pieces in all cases covered the same digits, the problem would have no solution; indeed, we would get 1 by subtracting two equal digits.

We determine first that the five-digit quotient is obtained in three steps. Thus, two of the five digits must be 0. It's obvious that these are the second and fourth digits—that is, both white knights cover zeros. Further, we see that if we multiply the two-digit divisor by 8, we obtain a two-digit result, therefore, the divisor cannot be greater than 12 (since $12 \cdot 8 = 96$). If we multiply the divisor by the first or the last digit of the quotient, however, the result is a three-digit number. The digits covered by the two white castles are thus greater than 8 and consequently both are 9.

So we've already derived the quotient: 90,809. We see that 12 is the only two-digit number that multiplied by 8 yields a two-digit product and multiplied by 9 yields a three-digit product. With the help of the divisor and the quotient, we can—using the assumption that the remainder was 1—also figure out the dividend:

$$1 + 90{,}809 \cdot 12 = 1{,}089{,}709.$$

11th Week

In the Garden

The answer is in the affirmative. He can start on the second stairway and return to the villa via the third. Or, he can start on the third stairway and return to the villa via the second.

The necessity of starting with these stairways can be demonstrated in exactly the same way as we argued in the solution of the puzzle "Nicholas's House" (3rd Week, p. 4): namely, the number of paths that meet at the second and third stairways is odd, and the number that meet at all other points is even.

Five Little Decks of Cards

Collect the decks of cards from the members of the audience so that the cards are face down and so that the order of the decks can be remembered. For example, put the deck of the first audience member (from the left) on the bottom, put the deck of the second audience member on top of that, and so on.

Now deal the cards out one at a time into five decks, and fan these out. If the decks are fanned so that the bottom card shifts to the left and the top card to the right, then obviously the card furthest to the left came from the deck that the first audience member (reading from the left) originally chose. The second card came from the second audience member's deck, and so on. If, then, after the cards of a deck are fanned out, the third audience member says the card he memorized is among them, his card must be the third card from the left. If two people speak up, the cards they memorized can be assigned to them in a similar manner.

A Clumsy Division

The third digit of the quotient is obviously 0; its first digit is not less than 4 (because a four-digit number minus a three-digit number is a two-digit number), while its last digit is certainly greater than 4 (because a one-digit number times a three-digit number produces a four-digit number). Since the quotient divided by 9 leaves the remainder 3, and none of the digits can be greater than 9, the sum of the digits of the quotient is one of the numbers 3, 12, 21, or 30. Therefore, 21 is the only possible value.

Accordingly, the sum of the first and last digits equals 17, which is to say that only 8 or 9 can be the first and the last digits. The quotient is thus either 8,409 or 9,408. In view of the subproducts, however, it's clear that the fourth digit of the quotient is greater than the first; therefore, the quotient is 8,409.

We know further that the divisor multiplied by eight is a three-digit number. This means that the divisor is greater than 100, but less than 125, and divided by 9, it leaves the remainder 7. There are only three such numbers, namely, 106, 115, and 124. The numbers 106 and 115 aren't eligible, though, since the dividend wouldn't be a seven-digit number: 124 produces the only seven-digit number: $8{,}409 \cdot 124 = 1{,}042{,}716$.

The division we seek is thus $1{,}042{,}716 \div 124 = 8{,}409$.

12th Week

Cycle Tour I

Let us say a player is in position H if H is the number he has named. Clearly, position 90, and any position greater than 90 but less than 100, is hopeless, since the opponent can then reach 100 in *one* step. For the same reason, position 89 is a winning one, since then our opponent is forced into a losing position; 78 is likewise a winning position, since no matter which number our opponent names, we can reach 89 in the next step and thereby reach a winning position. Moving backward in this fashion, we see that 67, 56, 45, 34, 23, 12, and 1 are winning positions. If we reach any one of these positions, then, assuming we play correctly, we will win. From this it is clear that, assuming a correct strategy, the starting player always wins. We see too that 1 is the only winning starting number. For example, if to this opening our opponent, adding 3, answers $3 + 1 = 4$, there is once more just a single winning option, namely, the addition of 8, since only in this way can we reach the winning position of 12. And so on.

More Jealous Husbands

Let's call the three husbands I, II, and III and their wives 1, 2, and 3, respectively. The crossing can be accomplished in the following way:

1. I and 1 cross over;
2. I returns alone with the boat;
3. 2 and 3 cross over;
4. 1 crosses back alone with the boat;
5. II and III cross over;

6. II and 2 cross back;
7. I and II cross over;
8. 3 crosses back;
9. 1 and 2 cross over;
10. 2 crosses back;
11. 2 and 3 cross over.

* * *

Note that if the problem involves only two jealous husbands (7th Week, "Jealous Husbands," p. 10), then five crossings suffice; but even if these husbands aren't jealous, they'll have to make five crossings. We can see that five crossings are necessary if we think about it in the following way: imagine a man who is the boatsman; he has to bring himself as well as three other people across, whereby he will have to cross back twice. Let's say I is the boatsman:

1. I and 1 cross over;
2. I returns with the boat;
3. I and 2 cross over;
4. I crosses back with the boat;
5. I and II cross over.

If there are three non-jealous husbands, however, then nine crossings suffice. In this case, the boatsman has to bring five people across to the other shore and, to accomplish this, must return four times to his starting point. So jealousy has a price!

Toy Soldiers—On the Double!

There are four integers between 0 and 36 that, divided by 9, leaves the remainder 5:

$$5, \quad 14, \quad 23, \quad 32.$$

Of these, however, only the number 23 fulfills the other condition that, divided by 4, it leaves the remainder 3.

Every natural number n can be represented in the form $n = 36k + r$, where k is a nonnegative integer (that is, 0 or a natural number) and r is one of the numbers 0, 1, 2, 3, ..., 35. (That means that dividing n by 36 yields the quotient k and the remainder r.)

The number 36 is divisible by 9 and 4. Therefore, the division of n by 4 or 9 yields the same remainder as the division of r by 4 or 9. Accordingly, n meets the conditions only if $r = 23$, that is, $n = 36k + 23$.

So if we set up the soldiers in 36 rows, 23 soldiers are left out.

(Note that to solve this problem, it wasn't necessary to determine the exact number of soldiers.)

13th Week

Big Problem—Chess Problem

Let us place the white rook on any square of the chess board. Then we cannot put the black rook on the same row or the same column as the white rook, otherwise, they could take each other. This excludes 15 squares, so there are $64 - 15 = 49$ squares on which we can place the black rook. So, for each of the 64 squares on which we can place the white rook, there are 49 squares on which we can place the black rook. Thus, there are $64 \cdot 49 = 3{,}136$ possible ways to set up the two rooks.

Cycle Tour II

Let x be the maximum that can be added to the number named by the opponent (x is between 5 and 9)—remember that x is also the starting number. As in the solution of "Cycle Tour I" (12th Week, p. 114), we get $100 - x - 1, 100 - 2x - 2, 100 - 3x - 3, \ldots$ for the winning numbers.

Accordingly, we get the smallest winning number by considering what the remainder of the division of 100 by $x + 1$ is; in the cases $x = 5$, $x = 6$, $x = 7$, $x = 8$, and $x = 9$, the remainders are 4, 2, 4, 1, and 0, respectively. If the starting player in the first four cases names the number 4, 2, 4, or 1, he wins, assuming he continues to play correctly. But if $x = 9$, then the beginning player's opponent can always reach the first winning number, namely, 10.

Therefore, Steve should pick nine so that the starting player should lose.

With Time and Patience, You'll Go Far

Besides the two expressions given as examples in the text, we know of the following additional 31:

$$100 = 75 + 24 + \frac{9}{18} + \frac{3}{6}$$

$$100 = 98 + 1 + \frac{27}{54} + \frac{3}{6}$$

$$100 = 1 + 93 + 5 + \frac{4}{28} + \frac{6}{7}$$

$$100 = 94 + 5 + \frac{38}{76} + \frac{1}{2}$$

$$100 = 1 + 95 + 3 + \frac{4}{28} + \frac{6}{7}$$

$$100 = 91 + 8 + \frac{27}{54} + \frac{3}{6}$$

$$100 = 95 + 4 + \frac{38}{76} + \frac{1}{2}$$

$$100 = 91 + 3 + 5 + \frac{4}{28} + \frac{6}{7}$$

$$100 = 57 + 42 + \frac{9}{18} + \frac{3}{6}$$

$$100 = 52 + 47 + \frac{9}{18} + \frac{3}{6}$$

$$100 = 3 + \frac{69{,}258}{714}$$

$$100 = 81 + \frac{5643}{297}$$

$$100 = 81 + \frac{7524}{396}$$

$$100 = 82 + \frac{3546}{197}$$

$$100 = 91 + \frac{5742}{638}$$

$$100 = 91 + \frac{5823}{647}$$

$$100 = 94 + \frac{1578}{263}$$

$$100 = 96 + \frac{2148}{537}$$

$$100 = 96 + \frac{1428}{357}$$

$$100 = 96 + \frac{1752}{438}$$

$$100 = 1 + 2 + 3 + 4 + 5 + 6 + 7 + 8 \cdot 9$$
$$100 = -1 \cdot 2 - 3 - 4 - 5 + 6 \cdot 7 + 8 \cdot 9$$
$$100 = 1 + 2 \cdot 3 + 4 \cdot 5 - 6 + 7 + 8 \cdot 9$$
$$100 = 1 + 2 \cdot 3 + 4 + 5 + 67 + 8 + 9$$
$$100 = 1 \cdot 2 + 34 + 56 + 7 - 8 + 9$$
$$100 = 12 + 3 - 4 + 5 + 67 + 8 + 9$$
$$100 = 12 - 3 - 4 + 5 - 6 + 7 + 89$$
$$100 = 123 + 45 - 67 + 8 - 9$$
$$100 = 123 - 45 - 67 + 89$$
$$100 = 123 - 4 - 5 - 6 - 7 + 8 - 9$$
$$100 = (1 + 2 - 3 - 4) \cdot (5 - 6 - 7 - 8 - 9)$$

The 33 examples given in this book are doubtless not all the possibilities by far. It's equally probable, though, that the number of the still-unknown expressions is limited and in total remains perhaps under 100.

(So I wrote 52 years ago in the first Hungarian edition, and did not change three years ago in the second Hungarian edition, nor last year in the German translation. How wrong I was. Here is an e-mail from a reader, Carsten Raudzis:

> I am reading your book, "Elmesport Egy Esztendőre" (or rather the German translation "Denksport für ein Jahr"). I really enjoy the puzzles and after solving them, I keep my colleagues, friends, and family occupied with them.
>
> However, I have a comment to one of the solutions. It is in Week 13, where you have to combine the numbers 1 to 9 with the basic arithmetic operations such that the result is 100. After finding 5 or 6 solutions more or less with educated guesses, I wondered how many solutions there are for this problem. Since my solutions are different from the ones you found, I assumed that probably there are many more than the 100 or so solutions you suggested.
>
> So, I wrote a little program which randomly sorts the numbers, randomly selects the operations and checks if the result is 100. To my great surprise, when I interrupted the program after several hours, it had saved more than 75,000 solutions.
>
> The program found a new, unique solution every few seconds. To estimate the total number of solutions, I proceeded as follows: I calculated the total number of possible expressions with any result. This is $N = 9! \cdot 5^8 = 1.42 \cdot 10^{11}$. Running

> my program several times I found an average of 1.5 solutions for every 10,000 randomly chosen calculations. So I estimate the number of solutions for this problem as $20 \cdot 10^6$ (order of magnitude)!
>
> I admit that this is only a rough estimate (after all, I am a physicist and not a mathematician...) but I thought you could be interested in this result nevertheless.

I posted on my Web site a list of 24,000 solutions by Carsten Raudzis.)

We see a very different picture if we also admit exponentiation. Consider the following examples:

$$100 = 94 + 5 + 678^{1+2-3},$$
$$100 = 97 + 2 + 1^{34,568},$$
$$100 = 98 + 3 - 1^{2,466}.$$

With the help of these examples, we can suggest hundreds of new examples.

Of course, the number of examples grows enormously again if we also admit the operation $\sqrt[n]{x}$.

14th Week

Pouring Judiciously

There are many possible solutions; here is one.

After each step, we'll always state how much wine is in each container; the first number refers to the 8-quart container, the second to the 5-quart container, and the third to the 3-quart container.

1. Fill the 5-quart container from the 8-quart container; 3 quarts remain in the 8-quart container.

   (3, 5, 0)

2. Fill the 3-quart container from the 5-quart container.

   (3, 2, 3)

3. Pour the contents of the 3-quart container into the 8-quart container.

   (6, 2, 0)

4. Pour the wine now in the 5-quart container into the 3-quart container.

   (6, 0, 2)

5. Fill the 5-quart container from the 8-quart container.

   (1, 5, 2)

6. Fill the 3-quart container from the 5-quart container.

   (1, 4, 3)

7. Pour the contents of the 3-quart container into the 8-quart container, whereby the problem is solved.

   (4, 4, 0)

Can you find other solutions?

The Barber

The words of the barber are clearly contradictory.

Let us divide the men of the town into two groups: those who do not shave themselves and those who do shave themselves.

Which group does the barber belongs to? If he shaves himself, then he shaves someone who shaves himself—contrary to what he stated. If, on the other hand, he does not shave himself, then he does not shave someone who doesn't shave himself—although he asserted that he shaves all such men.

* * *

The barber's puzzle may strike some as frivolous. In fact, it has a very serious mathematical background. It's an example of a so-called "logical paradox." (The great English mathematician and philosopher Bertrand Russell was the first to formulate this paradox in a somewhat different form.)

At the beginning of the twentieth century, paradoxes of this type made it necessary to provide an axiomatic foundation of set theory, a branch of mathematics. We cannot say much here about the mathematical significance of this paradox, it would take us too far afield. However, we describe one more paradox of a more mathematical nature.

Consider the set of all integers that can be described by a phrase containing at most 70 characters (letters and punctuation marks). Examples of such numbers are 16 ("sixteen"), 7 ("the fourth prime number"), 63 ("three times twenty one" or "sixty plus the second prime number").

Next consider the set of all those integers that *cannot be described* with at most 70 characters. A well-known theorem says that any nonempty set of natural numbers contains a *smallest* integer.

Consider now the following definition: "the *smallest* integer that *cannot be described* with 70 characters." Such an integer certainly exists: there cannot be more than 60^{70} such descriptions, since the English language does not have more than 60 characters. Among the integers that cannot be described with at most 70 characters, there is a uniquely determined smallest number that can be described as required.

The smallest non-describable number is then describable. Contradiction!

Council of Elders

(1) The *cook* is the father-in-law of the *baker*, and the daughter of Carter is married.

(2) Cook is engaged to the only daughter of the *carpenter* (who turned down proposals from the *smith* and the *potter*), and the daughter of Taylor is engaged.

(3) Since two members of the Council each have one daughter, *cook* is the profession of Carter, and *carpenter* is the profession of Taylor.

(4) Potter is a bachelor and his profession is not *potter*.

(5) The brother of the wife of Smith is the *potter*, and the *potter* is unmarried.

(6) The *gardener* and the *tailor* have married each other's sister.

(7) Smith can be neither the *gardener* nor the *tailor*.

We also deduce the following:

(8) Since the profession of Taylor is *carpenter*, the profession of Carpenter is x and the profession of X is *gardener* (here x and X are unknowns).

(9) Since the profession of Carter is *cook*, the profession of Cook is y and the profession of Y is *tailor* (here y and Y are unknowns).

We also know that Smith is married; Cook and Potter are not married; the *baker*, the *gardener*, and the *tailor* are married; the *smith* and the *potter* are unmarried. Consequently, Cook can be only the *smith*, *potter*, *mason*, or *carter*. He cannot, however, be the *carter*, since then, according to (9), the profession of Carter could only be *tailor*. According to (2), though, Cook can be neither the *smith* nor the *potter*. From this it follows that Cook must be the *mason* and also Mason must be the *tailor*.

Among the names still left, because of (7), the name of the *gardener* cannot be Smith, and because of (8), the names Gardener and Carpenter are also ruled out. The only possibility for the *gardener* is the name Baker. From this it follows because of (8), also, that Carpenter is the *baker*. Now only the professions of Smith, Gardener, and Potter remain in doubt, and the possible professions still available are *smith*, *potter*, and *carter*. But since Smith is married, his profession can be only *carter*; thus it follows from (4) that the profession of Potter is *smith*. Finally, for Gardener, the only remaining profession is *potter*.

In summary:

Name	Profession	Name	Profession
Potter	smith	Tailor	carpenter
Baker	gardener	Carter	cook
Smith	carter	Mason	tailor
Carpenter	baker	Gardener	potter
Cook	mason		

15th Week

The Story of Josephus Flavius

They had to position themselves at the 16th and the 31st places. We reach this solution most easily by writing out the numbers 1 to 41 in a circle, and, beginning with the first number, crossing out every third number. The last two remaining numbers indicate the two positions we seek.

In the first round, we cross out the numbers divisible by 3: 3, 6, 9, ..., 39. In the second round, we cross out the numbers 1, 5, 10, 14, 19, 23, 28, 32, 37, 41 and in the third round, the numbers 7, 13, 20, 26, 34, 40. In the fourth round, the numbers 8, 17, 29, 38 are crossed out; and continuing on, the numbers 11, 25, 2, 22, 4, 35 are crossed out with only two numbers remaining. Josephus Flavius and his friend must thus have positioned themselves at the 16th and the 31st places.

Car Trip

The truck arrived in Chicago 18 minutes = 0.3 hours later than the passenger car. The truck was 0.3 h · 42 mi/h = 12.6 mi from Chicago when the car arrived there. Each hour, the truck traveled 66 mi − 42 mi = 24 mi less—which is to say each minute, 0.4 mi less—than the car. The deficit of

12.6 mi is made up in a number of minutes equal to the number of times 0.4 mi goes into 12.6 mi:

$$\frac{12.6}{0.4} = 31.5.$$

Accordingly, the deficit of 12.6 mi is made up in a span of 31.5 minutes:

$$31.5 \text{ minutes} = \frac{31.5}{60} \text{ hours}.$$

In this time, the car traveled

$$\frac{31.5}{60} \text{ hours} \cdot 66 \text{ mi/h} = 34.65 \text{ mi}.$$

Thus, the car passed the truck 34.65 mi from Chicago.

* * *

This problem can also be solved with an equation.

Let x be the distance from the point the car passed the truck to Chicago. The passenger car has covered this distance in $x/66$ hours, while the truck required $x/42$ hours. This latter time period is 18 minutes = 0.3 hours longer than the former:

$$\frac{x}{66} = \frac{x}{42} - 0.3.$$

Multiplying both sides by $6 \cdot 7 \cdot 11$ gives the equation $7x = 11x - 138.6$. Therefore,

$$x = \frac{138.6}{4} = 34.65 \text{ mi}.$$

The Importance of Plain Speaking

Let x, y, and z be the ages of my friend's father, my friend, and his sister, respectively. Assume that u years ago the father was the same age as the ages at that time of my friend and his sister combined. So, we have

$$x + 6 = 3(y - u), \tag{1}$$

$$x - u = (y - u) + (z - u), \tag{2}$$

$$y = x - u, \tag{3}$$

$$x + 19 = 2z. \tag{4}$$

The left-hand side of (2) is the same as the right-hand side of (3), so the two other sides must be equal. Therefore,

$$y = (y - u) + (z - u).$$

From this it follows that

$$z = 2u.$$

Substituting this value into (4), we get

$$x = 4u - 19.$$

We subsitute this into (3) and obtain

$$y = 3u - 19.$$

If we substitute these values into (1), we are left with only the unknown u in the equation:

$$4u - 13 = 3(2u - 19) = 6u - 57.$$

From this we obtain

$$\begin{aligned} u &= 22, \\ x &= 4u - 19 = 69, \\ y &= 3u - 19 = 47, \\ z &= 2u = 44. \end{aligned}$$

* * *

With more complicated problems of this sort, it's useful to introduce many unknowns, which makes the conditions of the problem more clear, so it's easier to set up equations. For this puzzle, we were able to answer the question easily with the use of four unknowns. Most puzzles of this sort are even solvable with just one or two unknowns, but then it's more difficult to set up the equations. As an example, let's solve this puzzle using only two unknowns.

So, let x and y be, respectively, the ages of my friend and his sister at the time their father was $x + y$ years old. (In the previous discussion, that was u years ago.) The difference between my friend's father's age and that of my friend is thus y years. At present, my friend is the same age his father was at the time specified above. Therefore, my friend is currently $x + y$ years old, which means that since that time, y years have passed. Accordingly, my friend's father is currently $x + 2y$ years old, and his sister, $2y$ years old.

The puzzle tells us that

$$x + 2y + 6 = 3x,$$
$$x + 2y + 19 = 4y.$$

From the first equation, it follows that $x = y + 3$, which we substitute in the second equation:

$$y + 3 + 2y + 19 = 4y.$$

From this it follows that $y = 22$ and $x = 25$. The father is thus currently $x + 2y = 69$ years old; my friend is $x + y = 47$ years old, and his sister is $2y = 44$ years old.

16th Week

■■■■■■■■■■■■■■■■□□□□□□□□□□□□□□□□□□□□□□□□□□□□□□□□□□□□

The Fly

First, calculate (in hours) when the two teams meet. If we call this time t, then the racing cyclists cover a distance of $25t$ mi in this time, while the amateur cyclists cover a distance of $15t$ mi. At the time of their meeting, the two distances add up to 120 mi, therefore,

$$25t + 15t = 120.$$

Accordingly, $40t = 120$, and so $t = 3$ hours.

In 3 hours, the fly flies between the two teams at a speed of 30 mph. Thus, by the time of its heroic death, it has covered a distance of 90 miles.

* * *

Some seek a solution of this puzzle as follows:

> First, the distance the fly travels to meet the amateurs is calculated, then the distance back to the racing cyclists, and so on.

We know now that such complicated computations are not necessary.

Astonishing Trick

The first seven times, we can tap any seven numbers. The eighth time, we must tap the number 12, then 11, then 10, and so on.

The trick is explained as follows: Since the highest number my partner can think of is 12, he won't reach 20 at the first seven "taps." So it doesn't matter which numbers we tap the first seven times. At the 8th tap, my partner will reach 20 if the number 12 was chosen. Thus, the eighth tap must be on 12. At the ninth tap, my partner will reach 20 if the number 11 was chosen; thus, we must tap the ninth time on 11. The tenth tap must be on 10, and so on. In this way, we do indeed tap on the number my partner chose when he shouts 20.

Prize Question

Thirty students participated in the contest. If each student who solved a problem solved only one problem, we could find the number n of those who did not solve a single problem by subtracting the number of participants who solved the first, those who solved the second, and those who solved the third problems from the total number of participants. In our case this is the value:

$$30 - (20 + 16 + 10). \tag{1}$$

Now consider the number of students who solved two problems. They are counted twice in the number that is subtracted in (1). So we have to add the corresponding number to (1) once:

$$30 - (20 + 16 + 10) + (11 + 7 + 5). \tag{2}$$

But there were students in the contest who solved all three problems, and the number of these problem solvers appears three times in the number that is subtracted in (1). Therefore, we have to add their number twice to (1). But this number has been accounted for three times in the last parenthesis of (2). Our expression will be correct, then, only if we once more subtract the number of those who solved all three problems. In the mathematics contest, then,

$$n = 30 - (20 + 16 + 10) + (11 + 7 + 5) - 4 = 3$$

students solved no problem at all.

* * *

Let's next solve this problem manipulating sets.

Let U be the set of all contest participants. Further, let A, B, and C be the set of those students who solved the first, the second, and the third problems, respectively.

We know the following about the number of elements of the individual sets:

$$\begin{aligned} |U| &= 30, \\ |A| &= 20, \\ |B| &= 16, \\ |C| &= 10, \\ |A \cap B| &= 11, \\ |A \cap C| &= 7, \\ |B \cap C| &= 5, \\ |A \cap B \cap C| &= 4. \end{aligned}$$

The number of students who solved no problems is

$$|U| - |A \cup B \cup C|.$$

The number $|A \cup B \cup C|$ can be found as follows. The number of students who solved exactly three problems is

$$|A \cap B \cap C| = 4.$$

The number who solved exactly two problems is

$$\underbrace{|A \cap B| - |A \cap B \cap C|}_{11-4=7} + \underbrace{|A \cap C| - |A \cap B \cap C|}_{7-4=3} \tag{3}$$

$$+ \underbrace{|B \cap C| - |A \cap B \cap C|}_{5-4=1} = 11. \tag{4}$$

The number of those who solved only the first problem is

$$20 - (7 + 3 + 4) = 6.$$

The number of those who solved only the second problem is

$$16 - (7 + 1 + 4) = 4.$$

The number of those who solved only the third problem is

$$10 - (3 + 1 + 4) = 2.$$

Now we can already calculate that

$$|U| - |A \cup B \cup C| = 30 - (4 + 11 + 6 + 4 + 2) = 30 - 27 = 3.$$

Thus, 3 students solved no problem at all.

The solution can also be illustrated by a "Venn diagram":

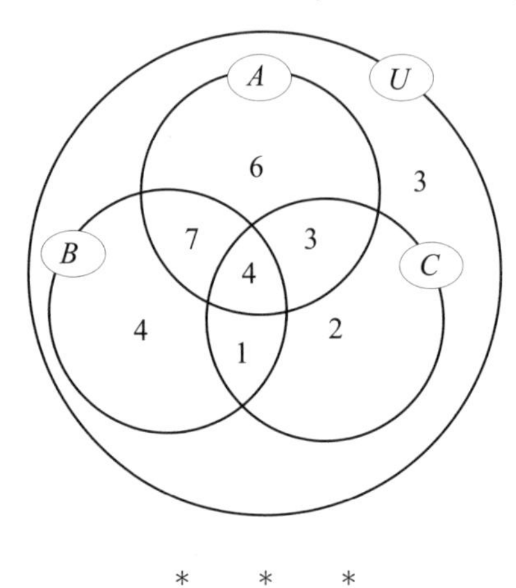

* * *

The puzzle can also be solved using the identity below, known as the *logical sieve formula*:

$$|A \cup B \cup C| = |A| + |B| + |C| - |A \cap B| - |A \cap C| - |B \cap C| + |A \cap B \cap C|.$$

This formula can be demonstrated by both of the methods we have employed. Using the logical sieve formula (also called the *inclusion-exclusion formula*), the solution of the problem goes as follows:

$$|U| - |A \cup B \cup C| = 30 - (20 + 16 + 10 - 11 - 7 - 5 + 4) = 3.$$

17th Week

Bicycle Motor

The additional distance a cyclist covers if he attaches a motor to his bicycle is

$$\frac{20\%}{40} = 0.5\%$$

of the total distance. The additional distance covered by 60% of the cyclists is equal to 150% of the total distance. From this it follows that 60% of the cyclists number

$$150\%/0.5\% = 300;$$

therefore, the total number of cyclists is 500. Accordingly,

$$500 \cdot 0.5\% = 250\%$$

is the additional distance covered if all 500 cyclists mount a motor on their bicycles. The total distance covered, then, increases by a factor of 3.5.

* * *

This puzzle, too, can easily be solved with equations. Let x be the factor by which the distance covered by *one* cyclist is increased by the addition of a motor. Further, let y be the number of cyclists. If 1 is the distance covered by *one* cyclist without a motor, then x is the additional distance covered with the aid of the motor; y is the total distance all the cyclists together can cover without a motor. Then,

$$y + 40x = 1.2y,$$
$$y + 0.6yx = 2.5y.$$

From the second equation, after dividing both sides by y, we obtain that $x = 2.5$, and so the first equation yields $y = 500$. (The values $x = 0$ and $y = 0$ also provide a solution of this system of equations, but this solution is meaningless for this problem.)

The additional distance covered is, then, 2.5 times the original combined distance, and thus the increased combined distance covered with the help of the motor is 3.5 times greater than the original distance.

Each Ninth Drops Out

For the simplest solution, once again, we draw a circle of the numbers, this time from 1 to 35. We cross off every ninth number, that is, the numbers

$$9,\ 18,\ 27,\ 1,\ 11,\ 21,\ 31,\ 6,\ 17,\ 29,\ 5,\ 19,\ 32,\ 10,\ 24,\ 3,\ 20.$$

Accordingly, he must position himself and his friends at the following places:

$$2,\ 4,\ 7,\ 8,\ 12,\ 13,\ 14,\ 15,\ 16,\ 22,\ 23,\ 25,\ 26,\ 28,\ 30,\ 33,\ 34,\ 35.$$

Impressive Card Trick

After the cards are laid out the first time, we place the deck that contains the memorized card in the middle. Then the cards of this deck occupy the places 10–18 in the combined deck. The memorized card, therefore, is one of the cards located at places 10–18.

Following the second laying out of the cards in three decks, we have (the numbers indicate the order following the first merging of the decks):

First deck:	1	4	7	<u>10</u>	<u>13</u>	<u>16</u>	19	22	25
Second deck:	2	5	8	<u>11</u>	<u>14</u>	<u>17</u>	20	23	26
Third deck:	3	6	9	<u>12</u>	<u>15</u>	<u>18</u>	21	24	27

The memorized card must be in a place indicated by one of the underscored numbers. We see that by knowing which deck contains the memorized card, we already know it is one of three cards whose places we marked. (For example, if the third deck is picked, then the number of the memorized card is 12, 15, or 18.) The third time we lay out the cards, the three possible cards end up in three different decks—in fact, in the center of their respective decks. Then we place the deck picked (the one with the memorized card) in the middle of the consolidated deck, and we find the card in the 14th place.

18th Week

How Did I Know That?

I knew it because if we start with any three-digit number that satisfies the condition, we always get 1,089.

To see this, let $100a+10b+c$ be any three-digit number with digits a, b, and c. Reversing the order of the digits, we form the number $100c+10b+a$. By assumption the two digits a and c differ by at least 2. Without loss of generality, we can assume that a is larger, so $100a + 10b + c$ is larger than $100c + 10b + a$. Thus,

$$100(a-c) + (c-a)$$

is the difference. Since $c - a$ is negative, it cannot serve as a digit.

Let us write this number in the decimal system:

$$100(a-c) + (c-a) = 100(a-c-1) + 10 \cdot 9 + (10 + c - a).$$

If we write the digits of this number in reverse order, we obtain the number

$$100(10 + c - a) + 10 \cdot 9 + (a - c - 1).$$

The sum of the two last numbers is

$$100(10-1) + 10 \cdot 18 + (10-1) = 1{,}089.$$

Let's see why it was necessary to assume that the first and last digits differ by at least two.

First, consider the case $a = c$. Then the number doesn't change by reversing the order of the digits. Hence, the difference is 0, and at the end of the procedure, we obtain 0, and not 1,089.

Next, let a and c differ by 1; let $a = c + 1$ (the case $c = a + 1$ is similar). Then we run into trouble when rewriting the number $100(a - c) + (c - a)$

in the decimal system, since $a - c - 1 = 0$. So, in this case, the difference is a two-digit number. (It still works if we view the two-digit difference as a three-digit number with 0 as the first digit, so the reversed number is a three-digit number ending in 0. For example, if we start with the number 786, then 687 is the reversed number, and the difference is 99. The sum of 99 and 990 is $1{,}089$.)

Weight Trick I

Let A, B, C, and D be the first four boxes, and let a, b, and c be the other three.

First, we put A, B, and C in one scale pan, and a, b, and c in the other. There are three possibilities:

1. If $ABC > abc$, then at the next weighing we put A and B on the two sides of the scale.
   (a) If $A > B$, then A is heavier than the other three boxes.
   (b) If $A < B$, then B is heavier than the other three boxes.
   (c) If $A = B$, then C is heavier than the other three boxes.
2. If $ABC < abc$, then at the next weighing we again put A and B on the two sides of the scale.
   (a) If $A > B$, then B is lighter than the other three boxes.
   (b) If $A < B$, then A is lighter than the other three boxes.
   (c) If $A = B$, then C is lighter than the other three boxes.
3. If $ABC = abc$, then at the next weighing, we put A and D on the two sides of the scale.
   (a) If $A > D$, then D is lighter than the other three boxes.
   (b) If $A < D$, then D is heavier than the other three boxes.

We see that two weighings always suffice, even though any box on the right might be lighter or heavier than the others.

Knight on the Chess Board

To land exactly once on every square, we must make a total of 63 moves to reach the upper-right corner. At every move, the knight lands on a square of the opposite color. At the start, it is on a white square, it lands at the first move on a black square, at the second move on a white square, at the third move on a black square, ..., at the 63rd move on a black square. The upper-right corner is a white square, however, so the knight cannot reach it at the 63rd move.

* * *

This puzzle is an interesting example of the mathematical analogue of the notorious principle *divide et impera* (divide and conquer). The often-employed mathematical principle states: "Divide up the properties of the object under investigation, and focus separately on these constituent elements, perhaps only on one or two of these elements."

Let's see how we might apply this to the solution of the present puzzle. The puzzle looks at first glance unfathomable. We begin by trying this and that, but sooner or later, all our attempts fail. Most people then lose interest, for who could try out every possible move just to prove the impossibility of the proposed task?

The trick is the use this principle; from among the various properties, we select just one (completely ignoring the complex geometry of a knight's move): the knight lands on a square of the opposite color at each move.

We can also generalize the statement of the puzzle: it's impossible to move from the lower-left square of the chess board to the upper-right square with a single piece in an odd number of moves such that with each move, the piece lands on a square of the opposite color.

One final comment: "This is all very well and good," someone might say, "but I can't see the use of all this subtlety." Just be patient! In one of the puzzles ahead, we'll see just how useful this little detour will prove.

19th Week

Pocket Change

If the result is an even number, then in his right hand he has an even number of coins; if the answer is an odd number, then he has an odd number of coins in his right hand.

The trick is based on the simple fact that the product of two even numbers, as well as the product of an even and an odd number, is always an even number, but that the product of two odd numbers yields an odd number. Further, the sum of an even number and an odd number is always an odd number, while the sum of two odd numbers, as well as the sum of two even numbers, is always even.

Water Pipes

It doesn't matter where Mr. Smith builds his well—he'll need the same length of pipe. Let A, B, and C be the corners of his yard, which is bounded by an equilateral triangle, and let P be any point in the garden. From P we draw a line perpendicular to each side of the triangle. Let P_A, P_B, and P_C be the points where these perpendicular lines meet the walls, respectively (as shown in the following illustration).

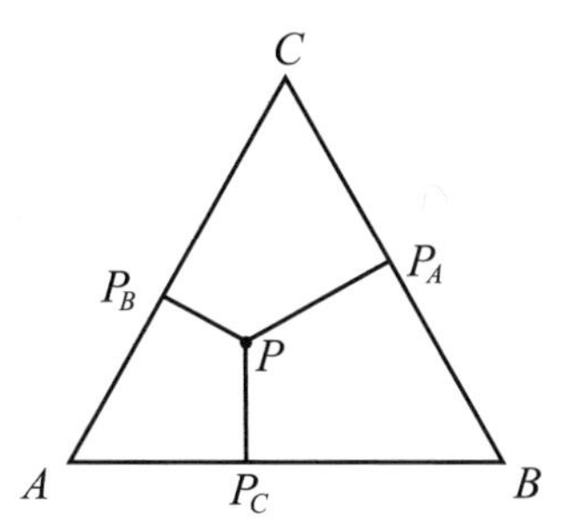

For any given point P, the required length of pipe will be minimal if the pipe is laid along the three perpendiculars, as shown in the illustration. But the value of the sum

$$PP_A + PP_B + PP_C$$

is independent of the choice of the point P. We can see this as follows.

The areas of the triangles PBC, PBA, and PAC, are given by

$$\frac{BC \cdot PP_A}{2}, \frac{BA \cdot PP_C}{2}, \frac{AC \cdot PP_B}{2},$$

respectively.

The sum of the areas of the triangles PBC, PBA, and PAC equals the area (denoted by f) of the triangle ABC. Using that $AB = BC = CA$, we get that

$$\frac{2}{AB} \cdot f = PP_A + PP_B + PP_C.$$

Since the left-hand side of the equation does not depend on the choice of P, the sum in question is independent of the placement of P in the triangle (including sides and corners).

Cycle Tour III

We can solve this puzzle arguing as in the previous cycle tour problems.

Clearly, 99 is the largest number which is a winning position, since if we say 99, our opponent is forced to say at least 100. From this number, we work our way backwards. Observe that whatever our opponent adds to the number he gets, we can always add enough to get 11 more than our previous number. For instance, if we say 88, then we can surely reach 99.

Thus, whoever says 88 is in a winning position; the same way, 77, 66, 55, 44, 33, 22, and 11 are also winning positions.

Thus, irrespective of what the starting player declares, his opponent can reach 11 and win, if he employs the correct strategy.

So Tommy was wrong; he would lose if Steve played skillfully.

20th Week

The Cans—Yes, We Can!

We can carry out the pouring procedure in 15 steps. (The numbers in parentheses give the quantity of wine in the 19-quart can, the 13-quart can, and the 7-quart can, respectively.)

1. Pour the contents of the 7-quart can into the 19-quart can.

   (7, 13, 0)

2. From the 13-quart can, fill the 19-quart can.

   (19, 1, 0)

3. From the 19-quart can, fill the 7-quart can.

   (12, 1, 7)

4. Pour the contents of the 7-quart can into the 13-quart can.

   (12, 8, 0)

5. From the 19-quart can, fill the 7-quart can.

   (5, 8, 7)

6. From the 7-quart can, fill the 13-quart can.

   (5, 13, 2)

7. Pour the contents of the 13-quart can into the 19-quart can.

   (18, 0, 2)

8. Pour the contents of the 7-quart can into the 13-quart can.

   (18, 2, 0)

9. From the 19-quart can, fill the 7-quart can.

   (11, 2, 7)

10. Pour the contents of the 7-quart can into the 13-quart can.

    $(11, 9, 0)$

11. Fill the 7-quart can from the 19-quart can.

    $(4, 9, 7)$

12. Pour the contents of the 7-quart can into the 13-quart can.

    $(4, 13, 3)$

13. Pour the contents of the 13-quart can into the 19-quart can.

    $(17, 0, 3)$

14. Pour the contents of the 7-quart can into the 13-quart can.

    $(17, 3, 0)$

15. Finally, from the 19-quart can, fill the 7-quart can.

    $(10, 3, 7)$

And now there are 10 quarts of wine in the 19-quart can.

Two Puzzles in Two Languages

We start with the English puzzle. If the sum of two four-digit numbers is a five-digit number, 1 is the only number that can be in the first place, so $M = 1$. The sum $S + M$ must be at least 9, and therefore, S is either 8 or 9. Thus, $S + M$ equals 9 or 10 and O is 0 or 1. And since 1 is no longer eligible, we have $O = 0$. The hundreds column can only yield a number to carry if $E = 9$ and $N = 0$, which is, however, not possible. So $S = 9$. The tens column must yield a number to carry, since otherwise we would have $E = N$. So $N = E + 1$. Accordingly, in the tens column it must be true that $N + R = E + 1 + R = E + 10$, if there is no number to carry in the ones column. But then we would have $R = 9$, which is not possible (since $S = 9$). So the ones column yields a number to carry, and we have $R = 8$. Thus, $D + E = 10 + Y$. And since Y can be neither 0 nor 1, $D + E$ is at least 12. D, on the other hand, is at most 7, and thus E is at least 5. Since N cannot be greater than 7, and $N = E + 1$, therefore, E can only be either 5 or 6. But 6 is not a possibility since then D and N would be equal to 7. This means that we have $E = 5$ and $N = 6$. So only $D = 7$ and $Y = 2$ are possible, and the solution is:

$$\begin{array}{r} 9567 \\ +\,1085 \\ \hline 10652 \end{array}$$

Next, the Hungarian puzzle. If the sum of two five-digit numbers is a six-digit number, 1 must be the first digit of the sum, so $E = 1$. Then $N + S$ can only be 11, since neither can be 1.

The sum $1+E+U = 2+U$ also yields a number to carry, since $J+J = 2J$ is an even number, and can, therefore, not end in 1. Accordingly, for $2J + 1$, we get either the value 1 or 11, that is, we have either $J = 0$ or $J = 5$.

Thus, $2+U = 10+J$, which gives us $U = 8+J$, excluding the possibility that $J = 5$ since U can't be 13. Thus, we have $J = 0$ and $U = 8$.

For the six remaining unknown letters, we have the following four equations, in which the unknowns must be different one-digit numbers that are not 0, 1, or 8:

$$N + S = 11, \tag{1}$$

$$L + Á = S, \tag{2}$$

$$É + M = 10 + L, \tag{3}$$

$$É + L + 0 + 1 + N = M + Á + 0 + 8 + S. \tag{4}$$

The last can be simplified to

$$É + L + N = M + Á + S + 7.$$

Note that we did not write (2) as

$$L + Á = 10 + S, \tag{2'}$$

because in that case, instead of equation (3), we would get the equation

$$É + M + 1 = 10 + L, \tag{3'}$$

and the addition of equations (1), $(2')$, and $(3')$ would yield

$$N + Á + É + M = 30;$$

this is not possible, since the left-hand side can be at most $9+7+6+5 = 27$.

From (2) and (3) it follows that

$$É + M - 10 + Á = S. \tag{5}$$

Thus, from (1), it follows that

$$N + É + M + Á = 21. \tag{6}$$

If we now substitute the value of L from (3) and the value of S from (5) into (4), we obtain that

$$\acute{E} + N - M - 2\acute{A} = 7. \tag{7}$$

Subtracting (7) from (6) yields

$$2M + 3\acute{A} = 14.$$

$\acute{A}$ must be even, and indeed must be either 2 or 4. But 4 isn't possible since then M would equal 1; consequently, $\acute{A} = 2$ and $M = 4$. From (2) and (5) follows that $\acute{E} - L = 6$. Since L is at least 3 (0, 1, and 2 are already taken) and $\acute{E}$ is at most 9, we obtain that $\acute{E} = 9$ and $L = 3$.

Now (5) and (6) give the values $S = 5$ and $N = 6$.

So the solution is:

$$\begin{array}{r} 93016 \\ +42085 \\ \hline 135101 \end{array}$$

Shopping for Chocolate

Let x be the number of chocolate bars purchased by one of the young men, as well as the number of one-dollar bills he spent. Similarly, let y be the number of chocolate bars purchased by his girlfriend, as well as the number of one-dollar bills she spent. Then

$$x^2 + y^2 = 65,$$

since anyone who bought x chocolate bars paid x dollars for each bar, and so paid altogether x^2 dollars. We reason similarly for y.

For the other couple, let u and v be the number of chocolate bars they purchased, so

$$u^2 + v^2 = 65.$$

Since x and y are natural numbers, the solutions are

$$\begin{array}{ll} x = 1, & y = 8, \\ x = 4, & y = 7, \\ x = 7, & y = 4, \\ x = 8, & y = 1. \end{array}$$

Similarly, we get,

$$\begin{array}{ll} u = 1, & v = 8, \\ u = 4, & v = 7, \\ u = 7, & v = 4, \\ u = 8, & v = 1. \end{array}$$

Since Betty bought 1 chocolate bar, her boyfriend bought 8 bars. Peter bought 1 bar more than Theresa. That can be the case only if Peter bought 8 and Theresa bought 7. Since $8^2 + 7^2$ is not 65, they are not a couple. Thus, Betty is Peter's girlfriend, and accordingly, Betty's boyfriend's name is Peter. (Paul is Theresa's boyfriend, and he bought 4 chocolate bars.)

21st Week

Instant Addition

The trick is to multiply the seventh number by 11; the product is the desired sum.

It's easy to verify this. Let a and b be the numbers selected by the friend. Then the sequence of numbers is as follows:

$$a,\ b,\ a+b,\ a+2b,\ 2a+3b,\ 3a+5b,\ 5a+8b,\ 8a+13b,\ 13a+21b,\ 21a+34b$$

Adding up these 10 numbers gives $55a + 88b$, and this number is in fact 11 times the seventh number, $5a + 8b$.

* * *

If you're good at doing quick calculations in your head, it's easy to figure out the original numbers. The simplest way is to subtract the tenth number from 7 times the sixth number:

$$7 \cdot (3a + 5b) - (21a + 34b) = b.$$

In this case, for example, we have $7 \cdot 49 - 338 = 343 - 338 = 5$, which is indeed the second number. Next, subtract five times the second number (which we now know) from the sixth number, and divide the result by 3. In our case, we obtain

$$49 - 5 \cdot 5 = 24,$$
$$24/3 = 8,$$

which is, indeed, the first number.

* * *

Here's how the sequence is formed: name two numbers, add them together, thus obtaining the third number; then add the second and third numbers, obtaining the fourth number, and so on. Such a sequence is called a *Fibonacci sequence*. Do you remember this sequence from Dan Brown's thriller, *The Da Vinci Code*?

The Fibonacci sequence is highly significant in number theory. If, for example, we begin the Fibonacci sequence with the numbers 1, 1, it has the remarkable property that the square of any number in the sequence differs by only 1 from the product of the numbers immediately preceding and following it. This sequence begins with the numbers

$$1, \quad 1, \quad 2, \quad 3, \quad 5, \quad 8, \quad 13, \quad 21, \quad 34, \ldots,$$

and then we have, for instance,

$$8^2 = 5 \cdot 13 - 1, \qquad 13^2 = 8 \cdot 21 + 1.$$

This property of the Fibonacci sequence is used in many puzzles. For instance, cut a chess board (8×8 squares) into four pieces and rearrange them into a 5×13 rectangle, which has 65 small squares:

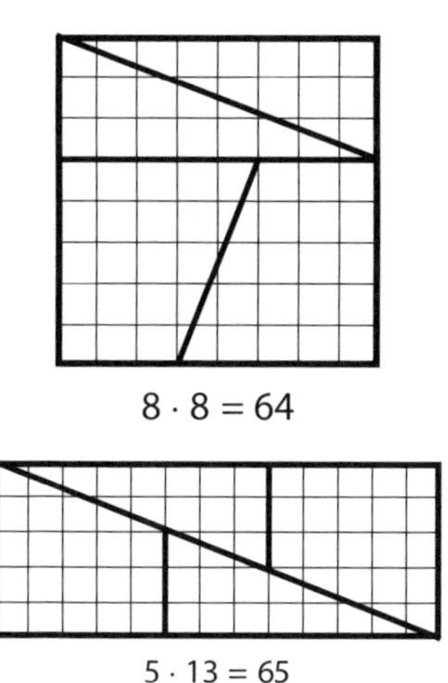

$8 \cdot 8 = 64$

$5 \cdot 13 = 65$

Magic! We gained a square!

The extra square slipped in by way of the long diagonal of the rectangle; the "squares" along it are not exactly square by a very small factor. Similarly, with any other member of the Fibonacci sequence, a square can be made to vanish or magically appear. For instance, from 13^2 (that is, from a 13×13 chess board), we can make an 8×21 rectangle:

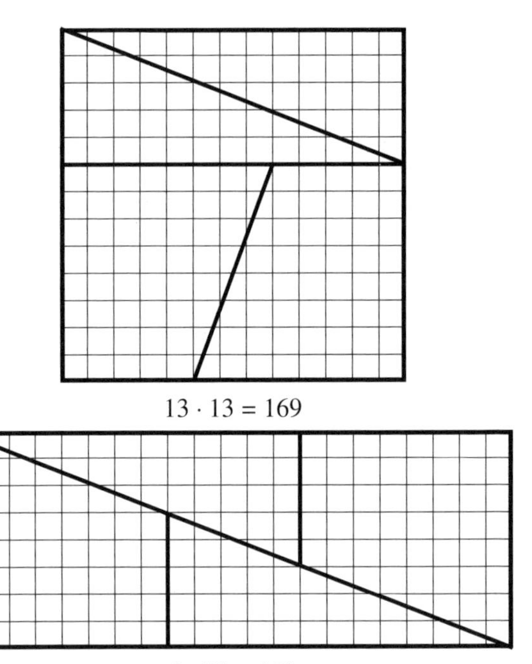

$13 \cdot 13 = 169$

$8 \cdot 21 = 168$

Magic! We lost a square!

Still More Jealous Husbands

On this excursion, the jealous husbands were not successful. Under these conditions, more than three married couples cannot be transported in a two-person boat to the opposite bank of the river. Why not?

Let n be the number of couples (so we have $2n$ participants), where n is at least 4. We call the riverbank they started from the near bank, and the opposite bank the far bank.

We proceed from two facts:

1. If there are both husbands and wives on a riverbank, the wives are a minority, or there are an equal number of husbands and wives.

   Indeed, otherwise one of the wives would not have her husband present, and would be in the company of other men.

2. If h is a number between 2 and $2n$, then in the course of crossing the river, a situation must arise in which there are h people on the far bank along with the boat.

   Indeed, after the first crossing, the number of people on the far bank, after the return of the boat and a new crossing, can change by at most 1.

Assume first that n is an even number, and consider the situation (which does happen by the second fact) when $n+1$ people and the boat are on the far bank of the river and there are $n-1$ people on the near bank. Since $n+1$ and $n-1$ are both odd, one of the banks must have more wives than husband, so by the first fact, there are no husbands on this bank. Therefore, all the husbands are on the opposite bank; specifically, on the other bank there are $n+1$ people (all the husbands and one wife), and of course the boat is also on that bank. The number n, however, is at least 4 and therefore there are at least three wives on this bank. This situation could arise only if one wife alone brought back the boat. And now only she can return with the boat, for no husband may cross over to the riverbank on which there are only wives (moreover at least three). There is no way out of this impasse.

Next assume that n is odd and consider the situation when there are n people on both riverbanks, and the boat is on the far bank. Reasoning as in the first case, we see there could only be husbands on one bank, and only wives on the other; otherwise, there would be more wives than husbands on one bank. If the husbands are on the far bank, then in keeping with the rules none of them can return to take wives over. But if the husbands are on the near bank, then they may not cross over, for irrespective of whether one husband crosses or two, there will be fewer husbands than wives on the other bank, since on the far bank there are at least three wives.

So both cases lead to contradiction. Since the two cases cover all possibilities, the transport scheme cannot be realized.

Soccer

Each team played three games.

Thus, the Revolution won its five points with two wins and one tie. (One win and three ties also yield five points, but requires four games.)

DC United had one win, against the Galaxy, and therefore, its three points were obtained with one win, one tie, and one loss. In view of the score $2:1$ against the Galaxy, the fact that DC United scored five goals, and also the fact that a total of 11 goals were scored, the following lists all eight possibilities for DC United's tie (left) and loss (right):

$$\begin{array}{ll}
0:0, & 3:4; \\
0:0, & 3:5; \\
1:1, & 2:3; \\
1:1, & 2:4; \\
2:2, & 1:2; \\
2:2, & 1:3; \\
3:3, & 0:1; \\
3:3, & 0:2.
\end{array}$$

The Revolution won two games. One of these victories must have been against DC United. If the game between the Revolution and DC United ended in a tie, the Revolution would have won both of its other games (scoring at least one goal in each), and in that case the total number of goals would be greater than 11.

The Revolution's other victory ended with the score $1 : 0$. The Revolution couldn't have scored more than one goal; otherwise there would again have been more than 11 goals in all. This one goal could only have been scored if in the games against DC United a total of at most ten goals were scored. Therefore, of the list of scores given above, we eliminated all but the following four:

$$\begin{array}{ll} 0:0, & 3:4; \\ 1:1, & 2:3; \\ 2:2, & 1:2; \\ 3:3, & 0:1. \end{array}$$

On the basis of our reasoning so far, and the fact that among the possible outcomes the total number of goals is 11, it is already clear that no more goals were scored. Thus, the match between the Wizards and the Galaxy was a scoreless tie, accounting for the Galaxy's one point. Accordingly, in the game against the Revolution, which ended with a score of $1 : 0$, the Galaxy must have suffered a loss.

The score between the Revolution and the Galaxy was $1 : 0$.

* * *

We see that the assumptions don't determine the results of all the games. Each of the four outcomes listed for the DC United games is possible, and leads to a different outcome for the tournament. Of course, the $1 : 0$ outcome of the game between the Revolution and the Galaxy is the same for all possibilities.

22nd Week

Prime Number

We prove first that every prime greater than 3 can be written in the form $6n + 1$ or $6n + 5$.

Let p be a prime number greater than 3; divide it by 6. Let n be the quotient and k the remainder, that is, $p = 6n + k$. The remainder k is 0, 1, 2, 3, 4, or 5.

If $n = 0$, then only $k = 4$ or $k = 5$ are possible since $p > 3$. If $k = 4$, then $p = 4$, which is not a prime. If $k = 5$, then $p = 5$ is a prime number, which can indeed be written in the form $6 \cdot 0 + 5$. So the statement is true for $n = 0$, and we can henceforth assume that n is not 0.

Numbers of the form $6n + 2 = 2(3n + 1)$ and $6n + 4 = 2(3n + 2)$ are divisible by 2, so they cannot be prime. Similarly, numbers of the form $6n + 3 = 3(2n + 1)$ are divisible by 3, so they cannot be prime. Of course, we can also discard $k = 0$, since then the number would be divisible by 6. Thus, only $k = 1$ and $k = 5$ are possible.

We have, then, proved that every prime number greater than 3 when divided by 6 leaves the remainder 1 or 5.

Now the problem is easily solved. Let p be the prime number selected. It can be written in the form $6n + 1$ or $6n + 5$.

Thus, after squaring, we get a number in the form $36n^2 + 12n + 1$ or of the form $36n^2 + 60n + 25$. If we add 17, we get $36n^2 + 12n + 18 = 12(3n^2 + n + 1) + 6$ or $36n^2 + 60n + 42 = 12(3n^2 + 5n + 3) + 6$. Dividing by 12, these two numbers obviously leave the remainder 6.

* * *

We can solve the problem in another way, too, based on our knowledge of algebra.

Let p be the prime number selected. First, square it, then add 17, to get $p^2 + 17$. We have to demonstrate that dividing this number by 12 leaves the remainder 6. Compute:

$$p^2 + 17 = (p^2 - 1) + 18 = (p + 1)(p - 1) + 18.$$

Since p is an odd number, it follows that both $p + 1$ and $p - 1$ are even numbers. Accordingly, the product $(p + 1)(p - 1)$ is divisible by 4. Since $p - 1$, p, and $p + 1$ are three consecutive integers, and p is a prime number greater than 3, it follows that either $p - 1$ or $p + 1$ is divisible by 3. So $(p+1)(p-1)$ is divisible by 12, and $(p+1)(p-1)+18$ divided by 12 leaves the remainder 6.

Cycle Tour IV

We can demonstrate that only $p = 10$ is a suitable value. If p is greater than 10, then Steve—if he plays correctly—will win every time. On the other hand, if p is less than 10, Tommy always wins.

Generally, we assume that one of the players, let's say Steve, may use the numbers from 1 to p—to start the game or to add to a number named by his opponent—and Tommy may use the numbers 1 to n similarly. Let us further assume that p is greater than n. We'll show that in this case, Steve always wins.

First, we show that if Steve achieves a winning position, Tommy can never stop him from winning. If Tommy adds any number r to the number Steve named (where r is strictly smaller than p), then Steve adds $p + 1 - r$ to the new number and again achieves a winning position. (We noted in earlier solutions of the Cycle Tour puzzles that if p is the largest number a player may use—to start or to add to a number named—then adjacent winning positions are $p + 1$ apart.)

If in any phase of the game Tommy is not in a winning position for Steve, then Steve can reach one of his winning positions in one move. But if Tommy is in a winning position for Steve, then Steve adds 1, and thereby the sum differs by p from Steve's next winning position. Tommy is then forced to add a number r that is smaller than p, and now Steve, by adding $p - r$, reaches a winning position.

The same argument holds for $n > p$; Tommy always wins. Therefore, only $p = n$ is fair; in this case, $10 = p = n$. But, we already learned from the solution of "Cycle Tour I" (p. 114) that in this case the starting player always wins.

Green Cross

Mr. Sharp Logic reasoned as follows that the cross on his forehead was green:

The cross on my forehead is either green or white. If the cross on my forehead is white, then the other two candidates, Mr. Able and Mr. Smart, see one white cross and a green cross. Since Mr. Able is smart, and sees that we all raised our hands, he would have reasoned as follows: "Why did Mr. Smart with a green cross on his forehead raise his hand? He raised his hand because he saw a green cross. But since the cross drawn

on the forehead of Mr. Logic is white, he must have observed the green cross on my forehead; therefore, there is a green cross on my forehead." Since Mr. Able is intelligent, he would have drawn this conclusion in a few moments, and cried out: "I know."

But Mr. Able did not say anything, and therefore, the cross drawn on my forehead is not white. Thus the cross on my forehead is green.

23rd Week

Money for Ice Cream

Assume the uncle distributed the dimes, as described, corresponding to the four digits of the square of the number a. First, we show that a is one of the numbers 38, 62, or 88.

A square number cannot end with two identical odd digits because the next to last digit of an odd square number is always even. To show this, let $10n + m$ be an odd number—that is, let m be odd. Then

$$(10n + m)^2 = 100n^2 + 20nm + m^2.$$

The next to last digit of this number is the last digit of the sum of the last digit of $2nm$ and the tens-place digit of m^2. The last digit of $2nm$ is even, since $2nm$ is even. On the other hand, m is one of the numbers $1, 3, 5, 7$, or 9, and the tens-place digits of the squares of these numbers are even (the digit 0 is also an even number—in the cases of 1^2 and 3^2, 0 is the tens-place digit).

Thus, only even square numbers can end with two identical digits. Square numbers that are even are divisible by 4, and a number composed of the last two digits of a number divisible by 4 is also divisible by 4 (because by subtracting from the number the two-digit number, the result is divisible by 100). A two-digit number with the same digits is of the form $11k$, and $11k$ is divisible by 4 only if k is divisible by 4. Also, k is the last digit of an even square number, and so, it can be only 0, 4, or 6. Since 6 is not divisible by 4, there are only two possibilities: $k = 0$ and $k = 4$. In the first case, the number ends in two zeros, and in the second case, in two fours.

To summarize our result: if the last two digits of a square number are the same, those digits are either 00 or 44.

The last two digits of the square number in the puzzle are the same, since the two biggest children received the same amount of money. We also see that a^2 doesn't end in 00, since all children got enough to buy ice cream.

There are only three two-digit numbers whose square is a four-digit number ending in 44: 38, 62, and 88.

The squares of these numbers are 1,444, 3,844, and 7,744, respectively. Of these numbers, only 3,844 has the property that the sum of its digits is 19, that is, only $a = 62$ is possible. And since 8 is the largest digit of 3,844, I was able to buy 8 scoops of ice cream with my money, at one dime apiece.

At Lake Michigan I

The position we are discussing:

> Each pile is a single match, except for one pile, which contains several matches

is very important. Let us call it a *One-Peak Position*.

So in a One-Peak Position, if the number of piles of matches is even, then Tommy must pick up the *entire* pile that has more than one match; if the number of piles of matches is odd, then he must pick up *all but one* of the matches from the pile with more than one match. In either case, an odd number of piles containing just one match each remain. Tommy's opponent, Steve, will thus always remove an odd numbered pile, and so the last pile too. Thus, Tommy wins.

Conclusion: in a One-Peak Position, the player who moves first wins provided that he plays correctly.

The Envious Cousin

I based my conclusion on the fact that if the last four digits of a square number are the same, then these four digits must all be 0.

If the last four digits of a square number are the same, then—as argued in the solution of the puzzle "Money for Ice Cream" (this week, p. 145)—this number either ends in four zeros or in four fours. In the latter case, the number is of the form

$$10{,}000d + 4{,}444.$$

One quarter of this is

$$100 \cdot 25d + 1{,}111.$$

This number is a square number, ending in 11, contradicting the statement that the last but one digit of a square number is even—also verified in the solution of the puzzle "Money for Ice Cream" (this week, p. 145). This shows that the number can end only in four zeros, and so my envious cousin received no money from our uncle.

24th Week

A Ship Sails By

Let x paces be the length of the ship, while its speed is k-times our walking speed. Note that if $k = 1$, then we couldn't measure the length of the ship by walking in the same direction; therefore, $k \neq 1$. We distinguish two cases: $k > 1$ and $k < 1$.

Case 1: $k > 1$. The ship is moving faster than we are. If we are walking in the same direction as the ship, it will catch up with us, so that we begin our calculation at the prow of the ship. When we have walked 200 paces, the ship will have traveled a distance of $200 \cdot k$ paces. Since in this time the ship has moved straight past us, the difference between the distances we and the ship have traveled is the length of the ship. This means that

$$x = 200 \cdot k - 200.$$

If we walk 40 paces in the direction opposite the ship's, the ship advances by $40 \cdot k$ paces. Since in this time the ship has moved straight past us, the sum of the two distances traveled equals the ship's length. Thus,

$$x = 40 \cdot k + 40.$$

From these two equations, we get

$$200 \cdot k - 200 = 40 \cdot k + 40,$$
$$160k = 240.$$

And so $k = \frac{3}{2}$ and $x = 100$ paces.

Case 2: $k < 1$. We are walking faster than the ship, so we catch up with it walking in the same direction. So we begin our calculation at the stern. In this case, the first equation becomes

$$x = 200 - 200 \cdot k.$$

The second equation remains unchanged:

$$x = 40 \cdot k + 40.$$

Now we get

$$200 - 200 \cdot k = 40 \cdot k + 40$$
$$240k = 160,$$

and so $k = \frac{2}{3}$ and $x = 66\frac{2}{3}$ paces.

Though it wasn't mentioned in the puzzle, Tommy was of course aware of whether he was dealing with Case 1 or with Case 2 and calculated accordingly.

At Lake Michigan II

Regardless how well Steve plays, he will always lose provided Tommy plays correctly. Let's designate the basic position as (1, 2, 3). This symbolizes that there are three piles of matches, consisting of 1, 2, and 3 matches, respectively.

We shall see that Tommy, independently of what Steve does, can force Steve to set him up in a One-Peek Position and thereby Tommy wins (see "At Lake Michigan I" (23rd Week, p. 146)).

Let's see what options Steve has:

- *Option 1.* He picks up either pile 2 or pile 3 in its entirety, that is, he creates either the position (1, 0, 3) or the position (1, 2, 0), and both are One-Peek Positions for Tommy.
- *Option 2.* He picks up either one match from pile 2 or two matches from pile 3, creating thereby the position (1, 1, 3) or the position (1, 2, 1), and those are again One-Peek Position for Tommy.
- *Option 3.* From the third pile, he picks up one match, creating the position (1, 2, 2). Then Tommy's proper response is to pick up pile 1, putting Steve in the (0, 2, 2) position. Then Steve picks up either one or two matches from either of the remaining piles. In the first case, he puts Tommy in One-Peek Position; in the second, Tommy picks up one of the two matches left on the table, and forces Steve to pick up the last match.
- *Option 4.* Steve picks up pile 1, creating the position (0, 2, 3). In this case, there is only one correct response for Tommy to make: he picks up one match from pile 3, leaving the position (0, 2, 2), considered in Option 3.

Dinner Guests

The first part of the problem reads:

> A teacher couple (husband and wife, both teachers) invited a doctor couple and an engineer couple to dinner. The blonde wife is busy in the kitchen; her husband arranges a chess board. The doorbell rings. The host asks two brown-haired and two black-haired guests to come in.

Based of this information, we conclude that of the four guests (two couples) two have brown hair and two have black hair, and the hostess is blonde.

We know further that

> We bid all farewell, and a red-haired member of the party sees us out.

Conclusion 1: The teacher host has red hair, and his wife is blonde.

Next we consider that

> the doctors and another member of the party never told the truth. We know, however, that the others told only the truth.

Using this statement, we know that the black-haired member of the party who answered our fifth question—identifying the spouse as a doctor—isn't a doctor, since the doctor couple on this evening have never told the truth. Further, the answer could have come only from someone who has not told the truth, for anyone else would not have a doctor as spouse. If we also use the fact that, by Conclusion 1, neither of the teacher couple has black hair, then we can state the following:

Conclusion 2: This evening, a black-haired engineer (along with the doctor and spouse) never told the truth, while everybody else has told the truth.

Based on Conclusion 2, we can say that the teacher couple always told the truth, and thus the male teacher answered our fourth question truthfully, and therefore:

Conclusion 3: This evening, the teacher husband has always told the truth, and his blonde wife is named Petra.

Accordingly, Petra also has not lied; she has answered our third question truthfully, so we know there are two women in the party with the same hair color. Since Petra is blonde, her husband is red-haired, and there are two brown-haired and two black-haired people. It follows that:

Conclusion 4: The other two women have the same hair color, and they have either black or brown hair.

Paul certainly did not tell the truth. Had he told the truth, he couldn't be a doctor (since the doctors lied) or a teacher (since the teacher wife's hair is not brown); thus, he could only be an engineer, and his wife has brown hair. Had he told the truth, however, his wife, by Conclusion 2, would have black hair, since he would be the lying, black-haired engineer. That is a contradiction. And so

Conclusion 5: This evening Paul never told the truth, and his wife does not have brown hair. And since he has not told the truth, he is not (by Conclusion 3) a teacher, and so his wife is not blonde. Thus, his wife has black hair.

By Conclusions 4 and 5, the doctor wife and the engineer wife have black hair. Further, the husbands of these women have brown hair, since there are only two black-haired persons in the party. Thus, Paul also has brown hair, and thus he also cannot be—by Conclusions 4 and 5—the lying, black-haired engineer. Rather, he can only be a lying doctor. Thus we have

Conclusion 6: Paul is a brown-haired doctor, and his wife has black hair.

Given this information, we see that the engineer wife has black hair, and her husband is brown-haired. Thus, by Conclusion 2, it is also true that the engineer wife, this evening, never told the truth, whereas her husband always told the truth. Our second question, then, was answered truthfully, that is, Dora's husband is named Thomas. By Conclusions 3 and 6, then, we conclude:

Conclusion 7: Thomas is the male engineer and has brown hair. His wife, Dora, is black-haired and this evening she never told the truth.

> After a few minutes, Stephen, Thomas, and Paul turn their attention to chess problems, while Dora, Ella, and Petra talk about their jobs.

Using this statement, we find all the information that is still missing:

Conclusion 8: The name of the teacher husband is Stephen, and the doctor wife is named Ella.

In summary, Stephen is a red-headed teacher; his blonde wife is named Petra. Thomas is a brown-haired engineer; his wife, Dora, has black hair. Paul is a doctor with brown hair, and his wife, Ella, has black hair.

25th Week

Beside/Below

The order was Ace, 3, King, 7, Queen, 4, Jack, 6, 10, 2, 9, 5, 8. We can convince ourselves of this most easily (as in the solution of "The Story of Josephus Flavius," 15th Week, p. 122) by laying out 13 cards that have been arranged in a predetermined order, in accordance with the rules. Then we look at the order, and draw the appropriate conclusions.

*　*　*

We can also solve the problem this way: first, imagine thirteen open slots in a row, numbered from one to thirteen. We lay the cards in descending order (Ace, King, Queen, Jack, 10, 9, 8, 7, 6, 5, 4, 3, 2) so that after laying down the first card in the first slot, we skip a slot and lay the next card in the second unoccupied slot (when we come to the end of the row, we start over, laying cards only in every second still-empty slot).

We illustrate the process by the following two tables:

after 7 steps:	1	2	3	4	5	6	7	8	9	10	11	12	13
	a		b		c		d		e		f		g

after 13 steps:	1	2	3	4	5	6	7	8	9	10	11	12	13
	a	l	b	h	c	k	d	i	e	m	f	j	g

The numbers in the first row of each table are the thirteen slot numbers, while the letters represent the descending order of the cards: a (Ace), b (King), c (Queen), d (Jack), e (10), f (9), g (8), h (7), i (6), j (5), k (4), l (3), m (2).

Lesson in Economy

Let s be the speed of delivery in miles per hour (mph). (Note that s must be greater than 0.) The delivery distance is $1{,}000$ miles, and the basic cost is $1{,}000$ pounds.

If $s > 10$ mph, the surcharge is

$$(s - 10) \cdot 1{,}000 \text{ shillings} = (s - 10) \cdot 50 \text{ pounds},$$

otherwise it is 0.

The delivery trip lasts

$$\frac{1{,}000}{s} \text{ hours.}$$

If the trip lasts longer than 20 hours, that is, if

$$\frac{1{,}000}{s} \text{ hours} > 20 \text{ hours}$$

and

$$s < 50 \text{ mph},$$

then the penalty comes to

$$\left(\frac{1{,}000}{s} - 20\right) \cdot 25 \text{ pounds;}$$

otherwise it is 0.

Assuming that $10 \leq s \leq 50$, the delivery cost is

$$1{,}000 + 50s - 500 + \frac{25{,}000}{s} - 500 \text{ pounds,}$$

or

$$50s + \frac{25{,}000}{s} \text{ pounds.}$$

For which s is the sum

$$50s + \frac{25{,}000}{s}$$

the least possible?

This sum is the arithmetic mean of $100s$ and $50{,}000/s$. Using the inequality between the arithmetic and the geometric means (see solution of "A Journey by Air" (1st Week, p. 90)), we obtain the following lower limit:

$$\frac{100s + \dfrac{50{,}000}{s}}{2} \geq \sqrt{100s \cdot \frac{50{,}000}{s}} = \sqrt{5{,}000{,}000} \approx 2{,}236.$$

The lower limit is reached if the two summands are equal:

$$100s = \frac{50{,}000}{s},$$

in which case

$$s = \sqrt{500} \approx 22.36 \text{ mph.}$$

We still need to check to see if there is the possibility of a cost-effective delivery in case we drop the condition $10 \leq s \leq 50$. If $s < 10$ mph, the cost is

$$1{,}000 + \frac{25{,}000}{s} - 500 = 500 + \frac{25{,}000}{s} > 500 + \frac{25{,}000}{10} = 3{,}000 \text{ pounds.}$$

On the other hand, if $s > 50$ mph, then the cost is

$$1{,}000 + 50s - 500 = 500 + 50s > 3{,}000 \text{ pounds.}$$

Thus,

$$\sqrt{5{,}000{,}000} \approx 2{,}236 \text{ pounds}$$

is the least possible delivery cost, and we can achieve it only if we deliver the goods at a speed of

$$\sqrt{500} \approx 22.36 \text{ mph.}$$

*   *   *

The following is the solution in the language of mathematics:

> Let $f(s)$ be the transportation cost at the delivery speed s. Then $s_0 = \sqrt{500}$ is the absolute minimum of $f(s)$.

What is the *absolute minimum* and what is the *absolute maximum*?

We'll clarify the matter with an illustrative example. On a New England vacation, we travel to the summit of Mount Washington. From its summit, clearly, all trails lead downwards. Thus, on the mountain summit in relation to our surroundings to a radius of, let's say, 20 mi, we have attained a maximal height. We say we have reached a *local maximum*.

It's not clear, though, that this is our highest reachable spot. In fact, we could reach a higher elevation in the continental United States by climbing Mount Whitney. But restricting our consideration to the mountains in New Hampshire, there is no higher spot we could reach than the summit of Mount Washington. Mount Washington is, then, not only the maximal height of its immediate neighborhood, but also the absolute maximum in New Hampshire.

To summarize, we can say the following about the relationship between the local and absolute maximums: the local maximum refers to a suitable neighborhood. The absolute maximum is also a local maximum, but it does not lose its maximal property if we consider more than its immediate neighborhood.

This example also shows clearly that it would not have sufficed to show only in the neighborhood [10, 50] that $f(s)$ has a minimum at $s_0 = \sqrt{500}$.

* * *

One more remark concerning the function $f(s)$: we saw that we could not, with just a single formula, describe the dependence of $f(s)$ on s: For $s < 10$, the value of $f(s)$ is given by a different formula than for $10 \leq s \leq 50$ and for $s > 50$.

It can't be claimed, however, that $f(s)$ is an "artificial" function, since, after all, it has arisen in a practical problem. This example can also help dispel the notion that a "proper" functional relation is always given by a single formula.

Auto Racing

Four minutes are required to travel two miles at a speed of 30 mph. The driver traveled one mile at a speed of 15 mph. Therefore, he used the whole four minutes for just the first mile. It would have been possible to have arrived in Northy only a few seconds late only if the driver plunged off the top of the mountain.

26th Week

Uncle Pops a Quiz

Let t, u, v, x, y, and z be the numbers of workers in the six groups. Then

$$\begin{aligned} t(u+v+x+y+z) &= 264, \\ u(t+v+x+y+z) &= 325, \\ v(t+u+x+y+z) &= 549, \\ x(t+u+v+y+z) &= 825, \\ y(t+u+v+x+z) &= 1{,}000. \end{aligned}$$

We have to proceed from the basic fact that the sum of the two factors of the multiplication on the left-hand side is always the same: it equals the total number of workers in all six workers' groups, that is, $t+u+v+x+y+z$.

Accordingly, the numbers on the right-hand side have to be factored into two factors so that the sums of these pairs are the same. To do this, we first write out all possible factorizations of these integers into two integers:

$$\begin{aligned} 264 &= 1\cdot 264 = 2\cdot 132 = 3\cdot 88 = \underline{4\cdot 66} = 6\cdot 44 = 8\cdot 33 \\ &= 11\cdot 24 = 12\cdot 22 \\ 325 &= 1\cdot 325 = \underline{5\cdot 65} = 13\cdot 25 \\ 549 &= 1\cdot 549 = 3\cdot 183 = \underline{9\cdot 61} \\ 825 &= 1\cdot 825 = 3\cdot 275 = 5\cdot 165 = 11\cdot 75 = \underline{15\cdot 55} = 25\cdot 33 \\ 1{,}000 &= 1\cdot 1{,}000 = 2\cdot 500 = 4\cdot 250 = 5\cdot 200 = 8\cdot 125 = 10\cdot 100 \\ &= \underline{20\cdot 50} = 25\cdot 40 \end{aligned}$$

It is easy to see that only the underscored factorizations have the property that the sum of the factors is the same (70). From this, we quickly learn the number of workers in the groups:

$$t=4, \quad u=5, \quad v=9, \quad x=15, \quad y=20.$$

We get the value of z, for example, from

$$66 = u+v+x+y+z,$$

which yields $z=17$.

The Nephew Strikes Back

Using the idea of the solution of the previous puzzle, the answer is quite simple. If we look carefully at the factorizations of $1{,}000$ and 264 into two integers, we see that there is only *one* factorization for these integers with the property that the sum of the factors of $1{,}000$ is the same as the sum of the factors of 264.

In the factorization of 264 the sums of the factors are

$$265,\ 134,\ 91,\ \underline{70},\ 50,\ 41,\ 35,\ 34.$$

In the factorization of $1{,}000$ the sums of the factors are

$$1{,}001,\ 502,\ 254,\ 205,\ 133,\ 110,\ \underline{70}\,,\ 65.$$

The only number in common is 70; that, consequently, is the number of workers in the six groups.

The Window Dresser's Error

Let $\overline{txyz}$ be the originally marked price (overscoring indicates the four-digit number with the digits t, x, y, z).

The correct price is, then, $\overline{zyxt}$, and both numbers are squares. We also know that the correct price is an integer multiple of the price originally marked in the window display. This can be the case, however, only if the quotient of the two numbers is itself square. Considering also that the quotient of two four-digit numbers can only be a one-digit number, we see that the correct price is either 1 times, 4 times, or 9 times the originally marked price. Note that 1 times the marked price makes no sense in the context.

In case

$$\overline{zyxt} = 4 \cdot \overline{txyz},$$

there is an even square number on the left-hand side, which can thus end only in 0, 4, or 6. But if t is greater than 2, then the product on the right-hand side of the equation, $4 \cdot \overline{txyz}$, is a five-digit number. So for t, only 0 is possible. But then $\overline{txyz}$ would be a three-digit number. Accordingly,

$$\overline{zyxt} = 9 \cdot \overline{txyz}.$$

We know that $\overline{zyxt}$ is a four-digit number, and therefore, smaller than $10{,}000$. Thus, we have

$$\overline{txyz} < \frac{10{,}000}{9};$$

and so,

$$1{,}000 \leq \overline{txyz} \leq 1{,}111.$$

Between $1{,}000$ and $1{,}111$, there are only the squares of the numbers 32 and 33:

- $32^2 = 1{,}024$, but this provides no solution, since $4{,}201$ is not a square number.
- $33^2 = 1{,}089$ fulfills the conditions, since $9{,}801 = 99^2 = 9 \cdot 1{,}089$.

The price of the hat, then, was $\$9{,}801$.

27th Week

At Lake Michigan III

If there is one match in each of the two piles, then clearly the starting player wins.

But if in both piles the number of matches agree and is greater than 1, then the starting player loses, provided his opponent plays correctly. The player going second can win using the following strategy. He picks up matches from the pile his opponent did not touch so as to equalize the number of matches in the two piles. There are two exceptions: 1. The opponent picks up all the matches but one in one of the piles; then the player going second picks up the entire second pile. 2. The opponent picks up an entire pile; then the player going second picks up all but one match in the other pile.

This method can be employed until the end of the game. At each move, the first player is caught in a position from which there is no escape. If the second player makes a single error, however, the first player can utilize the same strategy to reach a winning position.

So, if there is more than one match in each of the two piles, and if both piles contain the *same number* of matches, the beginning player loses, provided the player going second plays correctly. But if there are a *different* number of matches in the two piles, then, assuming he plays correctly, the beginning player wins.

Colored Dice

We paint one face of the cube any color. Then we have five colors from which to choose to paint the opposite face. This leaves a ring of four faces. Again, we pick one face and paint it with one of the four remaining colors.

The remaining three faces can be painted in $3! = 6$ different ways. Thus, there are $6 \cdot 5 = 30$ different ways of painting the dice.

We can put the three number pairs (1, 6; 2, 5; 3, 4) on three pairs of faces in 3! ways on one painted die, but the digits of the number pairs can also be interchanged within the number pair, which leads in all cases to two new possibilities. Thus, the numbers can be put on the colored dice in

$$3! \cdot 2^3 = 6 \cdot 8 = 48$$

different ways.

Consequently, the total number of dice that can be made up in this way is $30 \cdot 48 = 1{,}440$. Each child has painted 20 dice, all different from one another. Therefore, there can be at most 72 children, since $72 \cdot 20 = 1{,}440$.

Francesca's Teachers

From (3) it follows that Ms. Evans does not teach mathematics. From (5) we see that Mr. Knight is the eldest, and he can be neither the physics teacher nor the biology teacher. It follows further that the biology teacher is the youngest, and that she doesn't teach physics.

By (2), we know that Ms. Anderson teaches biology, but not physics.

Since neither Mr. Knight nor Ms. Anderson teaches it, physics must be one of Ms. Evans's subjects.

According to (6) and (4), Mr. Knight teaches neither chemistry nor mathematics.

So, Mr. Knight teaches English and history. Then, Ms. Evans's other subject can only be chemistry. Ms. Anderson's second subject is then mathematics.

In summary, Ms. Anderson teaches biology and mathematics; Mr. Knight teaches English and history; and Ms. Evans teaches physics and chemistry.

28th Week

Scribble

Nor will we succeed, since there are a total of four points on the illustration at which an odd number of lines meet. As we have already seen in the puzzle "Nicholas's House" (3rd Week, solution on p. 92), every such point must be either a starting or an ending point.

It is clear, however, that if we are to draw a figure without lifting the pencil, there can be only one starting point and one ending point. (Note that these two can be the same, in which case, there is no point at which an odd number of lines meet.) Thus, this figure cannot be drawn.

Randolini's Card Trick

Randolini knows the order of the cards in the deck he hands out (it's of course enough to know the cyclical order of the cards—see the hint on p. 80). It's easy to see that cutting the deck doesn't change the cyclical order. We can lay the cards out in a circle, for example, and note which cards lie where; then we push the cards together again, cut the deck several times, and lay the cards out again. We see that the same picture presents itself, only rotated.

Now providing a method for finding the restacked card is easy.

We order the cards returned to us in their original cyclical order. There are two possibilities. Either there is one card that doesn't fit into the cycle, or there is a card missing in the cycle. In the first case, we've been handed back the stack that contains the card moved from the other stack, and this card doesn't fit into the cycle. In the second case, we've been handed the stack from which a card has been removed, and this is the card that is missing from the cyclical order.

Let's consider an example.

For simplicity's sake, let's assume there are only 13 cards in the deck: Ace (1), 2, 3, 4, 5, 6, 7, 8, 9, 10, Jack (J), Queen (Q), and King (K), in that order. After one cut, for example, the order could be 9, 10, J, Q, K, 1, 2, 3, 4, 5, 6, 7, 8, and after another cut, 6, 7, 8, 9, 10, J, Q, K, 1, 2, 3, 4, 5. (Note that any number of cuts can be replaced by a single cut.) The deck is now cut as follows into two small decks: one small deck contains 6, 7, 8, 9, 10, J, and Q, and the other, K, 1, 2, 3, 4, and 5. Let's say the 2 has been removed from the second small deck and put into the first small deck. Then the first small deck is shuffled and we now have the cards in the following order: 6, Q, 2, 9, 7, J, 10, 8. We picture the cards rearranged in their original cyclical order and recognize that the card 2 doesn't belong here. So that's the moved card.

Of course, in presenting this trick, it's essential that we use more cards than in the example and we start with a cyclical order we can memorize, but appears well shuffled to the audience.

Vacation

We start with the following table:

Name	Residence	Vacation City
	Memphis	
×	Dallas	Jackson
	Orlando	
	Phoenix	×
	Jackson	

The table contains the information that my friend who lives in Dallas took his vacation in Jackson, and that my friend who lives in Phoenix took his vacation in the city bearing the name of my friend from Dallas.

What is the name of my friend from Dallas? Clearly, he can't be named Dallas, since Dallas doesn't live in Dallas. He also can't be named Jackson, since he took his vacation in Jackson, and we know that Jackson spent his vacation in Phoenix. He can't be named Phoenix, because then my friend from Phoenix would have taken his vacation in Phoenix, since my friend who lives in Phoenix spent his vacation in the city that is named as my friend who lives in Dallas.

Nor can he be named Orlando, since Orlando took his vacation in the city which bears the name of my friend from Phoenix; in that case, the name of my friend from Phoenix would, however, be Jackson, who would have gone to Phoenix for his vacation, which we know is not possible, since my friend from Phoenix could not have taken his vacation in Phoenix. Accordingly, my friend from Dallas is named Memphis.

Who lives in Phoenix? It can't be Orlando; if it were, Orlando would have taken his vacation in Orlando. Of course he also can't be named Phoenix. And he also can't be named Memphis, because we already know that Memphis lives in Dallas. Nor can he be named Jackson, since then (as noted) my friend from Phoenix would have taken his vacation in Phoenix. The name of my friend from Phoenix is, thus, Dallas, and he traveled to Memphis on his vacation.

Who traveled to Orlando? The name of my friend who traveled to Orlando cannot be Orlando, and on the basis of our previous conclusions, it can't be either Memphis or Dallas. Nor can his name be Jackson, since that person traveled to Phoenix. Thus, Phoenix traveled to Orlando.

We don't know whether Phoenix lives in Memphis or Jackson. In either case, all the conditions are satisfied. Both solutions are given below:

Name	Residence	Vacation City
Phoenix	Memphis	Orlando
Memphis	Dallas	Jackson
Jackson	Orlando	Phoenix
Dallas	Phoenix	Memphis
Orlando	Jackson	Dallas

Name	Residence	Vacation City
Orlando	Memphis	Dallas
Memphis	Dallas	Jackson
Jackson	Orlando	Phoenix
Dallas	Phoenix	Memphis
Phoenix	Jackson	Orlando

29th Week

A Typical Mathematician

There is only one square number between 1800 and 1900:

$$43^2 = 1{,}849.$$

Note that $42^2 = 1{,}764$ and $44^2 = 1{,}936$.

So De Morgan turned 43 in 1849, and thus he was born in 1806.

Weight Trick II

In this case, also, two weighings will suffice. Divide the nine boxes into three groups of three boxes. Let A, B, and C be the three groups. First, we weigh A and B. If $A < B$, then A contains the lighter box. If $A > B$, then B contains the lighter box. If, finally, $A = B$, then C contains the lighter box. Thus, already the first weighing allows us to determine the group containing the lighter box.

Let a, b, and c designate the three boxes in the group we found; next we weigh a and b. If $a < b$, then a is the lighter box; if $a > b$, then b is the lighter box; and if $a = b$, then c is the lighter box.

* * *

We could have solved the problem similarly had the box in question been heavier, instead of lighter, than the others.

If we have 3^n boxes, we could utilize this method to find with n weighings the box whose weight was different. We form three groups, each group consisting of 3^{n-1} boxes. Once we've determined in which group the lighter (or heavier) box is located, we form within this group another three groups; each of these consists of 3^{n-2} boxes, and so on.

For example, we can find among $243 = 3^5$ boxes the one lighter box; we first form three groups, each consisting of 81 boxes. Then we take the group containing the lighter box and divide it up into three groups of nine. Next, we choose from among those three groups, each containing three boxes, that group whose weight is lighter, and finally, we determine which of the three boxes in that group is lighter.

If the number of boxes is between 3^{n-1} and 3^n, then we require likewise n weighings.

At Lake Michigan IV

Among the various match problems we've encountered so far, it's worth having a closer look at the solution of the puzzle "At Lake Michigan III" (27th Week, p. 156). There, we investigated the position in which there are two piles of matches. The winning strategy described there (*evening out the matches*) is a special case of the Binary Strategy.

Let's see how this strategy can be used in the present, more complicated, case.

Let's designate the four groups as position (6, 5, 5, 3), meaning that we have four groups of matches and that there are six matches in the first group, five matches each in the second and third groups, and three matches in the fourth group.

As required by the Binary Strategy, write 6, 5, 5, and 3 as binary numbers (the fundamental concepts of number systems can be found in the solution of the puzzle "Curious Addition" (9th Week, p. 106)), and then add up the numbers within the individual columns as decimal numbers:

$$\begin{array}{r} 110 \\ 101 \\ 101 \\ 11 \\ \hline 323 \end{array}$$

We see that there is an odd number in the first and third columns. If, for example, we take away $2^2 - 2^0 = 3$ matches from the first group, getting to the position $(3, 5, 5, 3)$, then the sums in the columns are even:

$$\begin{array}{r} 11 \\ 101 \\ 101 \\ 11 \\ \hline 224 \end{array}$$

We assume that our opponent now takes away four matches from the second group, obtaining the position (3, 1, 5, 3). Then summing the columns we get:

$$\begin{array}{r} 11 \\ 1 \\ 101 \\ 11 \\ \hline 124 \end{array}$$

So, for a winning position, we must remove 2^2 matches from the third group, leaving the position (3, 1, 1, 3). This will give us 24, after adding up the columns. Assuming our opponent removes the entire first group, then what remains is the position (0, 1, 1, 3), a Single-Peak Position.

* * *

In this explanation, we haven't proven the correctness of the strategy; we've only demonstrated how it can be used in practice.

Note that we could also have written the sums obtained from the columns of the aligned binary numbers as binary numbers, because whether a number is even or odd is independent of the number system in which the number is represented. Our recommendation was made just for convenience. If you want to calculate in the binary system, you may do so. We could have chosen the base-12 system!

30th Week

Carton Puzzles

In place of each letter, write the number of ways of reaching it. There is clearly only one way to reach the outside letters of the upper seven rows. At every other place, we obtain the appropriate number by adding the two numbers (diagonally, to the left and to the right) directly above it, since it is obvious that as many paths lead to a letter as to the two letters above it combined (because each letter can be reached from the two letters above it, and only from them):

1

1 1

1 2 1

1 3 3 1

1 4 6 4 1

1 5 10 10 5 1

1 6 15 20 15 6 1

7 21 35 35 21 7

28 56 70 56 28

84 126 126 84

210 252 210

462 462

924

Thus, there are 924 ways to read the phrase CARTON PUZZLES.

* * *

The question arises whether it is necessary to set up this table in order to solve the puzzle. In the solution of the puzzle "Say How Many Flags Are Flying" (6th Week, p. 99), we've already encountered the notation

$$\binom{n}{k} = \frac{n!}{k!(n-k)!},$$

which gives the number of combinations of k elements from a set of size n.

It is easy to see that in the kth place of the nth row, we find the number

$$\binom{n-1}{k-1}.$$

Thus, the question posed in the puzzle can be answered with a simple calculation, for the number sought is the 7th number in the 13th row:

$$\binom{13-1}{7-1} = \binom{12}{6} = \frac{12 \cdot 11 \cdot 10 \cdot 9 \cdot 8 \cdot 7}{2 \cdot 3 \cdot 4 \cdot 5 \cdot 6} = 924.$$

How can we explain this? Let's think about what it means to reach any letter when tracing the words CARTON PUZZLES along any given path. To complete a trace, 12 steps must be taken, and each step must be made to the left or the right. The chosen path might, for example, look like this: we go three steps to the right, then six steps to the left, and then three steps to the right. All paths must take a total of six steps to the right and six steps to the left. The number of possible paths corresponds, then, to how many ways we can select 6 elements out of 12 elements—in other words, how many ways we can select the steps taken to the right. (Here, the order of the selected elements is of no concern; it's all the same whether I say "At the first, third, fifth, sixth, eighth, and eleventh steps, I go right," or I say "At the fifth, sixth, third, eleventh, eighth, and first steps I go right.") Accordingly, in this case we have

$$\binom{12}{6}$$

possible paths. The table above is part of an infinite triangle in which at the $(k+1)$th place of the $(n+1)$th row is the number

$$\binom{n}{k}.$$

This triangle is called *Pascal's triangle*. Pascal's triangle has many intriguing properties. We've already discussed one such property: every "inner" element is the sum of the two numbers above it.

Here is another property: we can see, for example, that the coefficients in the right-hand side of the equation below—1, 3, 3, 1—appear in the fourth row:

$$(a+b)^3 = a^3 + 3a^2b + 3ab^2 + b^3.$$

In general, in the $(n+1)$th row we find those coefficients that appear in the expansion of the binomial expression $(a+b)^n$. Thus, these numbers are called *binomial coefficients*.

No Splitting of Hairs!

There are obviously two people who have exactly the same number of hairs, since bald people satisfy the condition. Ignoring this, however, we'll prove that among non-bald people too, there are at least two who have exactly the same number of hairs.

A human being has at most 150,000 hairs. Now imagine $150{,}000 + 1$ boxes. In the extra box, we place the bald people; in the first box, those who have one single hair; in the second, those who have two hairs; and so on.

Since every human being has at most 150,000 hairs, there is a box for everyone. What would it mean if the statement were false, that is, if there were no two people with the same number of hairs? That would mean that there is no box with two people; in other words, every box would contain either no people or one person. Thus, there would be at most 150,001 people in the boxes, and that would mean that at most 150,001 people live in Salt Lake City. We know, though, that at least 1.7 million people live in Salt Lake City. So, from the assumption that no two people have exactly the same number of hairs, we run into a contradiction.

By the same process, we also see that there must be at least 12 people living in Salt Lake City with the same number of hairs. Indeed, if there are

at most 11 people with the same number of hairs, Salt Lake City couldn't have more than $11 \cdot 150{,}001 = 1{,}650{,}011$ inhabitants. But since more people than that live in Salt Lake City, there must be at least 12 people with the same number of hairs.

*   *   *

The extraordinarily simple idea we employed is known as the *pigeonhole principle* or *drawer principle*. Though it sounds simplistic, it is in truth a very useful tool. The German mathematician Dirichlet employed this principle extensively; consequently, it is also known as *Dirichlet's box principle*. This is a simple principle whose use leads to very interesting and altogether nontrivial results.

Toy Soldiers—Halt!

Let n be the number of soldiers.

At first, $n - 5$ soldiers are set up in 24 even rows; the second time, $n - 16$ soldiers in 15 even rows. Therefore, $n - 5$ is divisible by 24, and $n - 16$ by 15. A number divisible by 24 and a number divisible by 15 are of course both divisible by 3. Therefore, $n - 5$ and $n - 16$ are divisible by 3. The difference between two numbers divisible by three is itself divisible by 3, and therefore,

$$(n - 5) - (n - 16) = n - 5 - n + 16 = 11$$

must also be divisible by 3. But 11 is not divisible by 3. This contradiction shows that the rows were not even.

*   *   *

In the solution of the earlier problem "Knight on the Chess Board" (18th Week, p. 131), we dealt with the mathematical counterpart of the *divide et impera* principle. Here, we see another example of its applicability. In solving this puzzle, we didn't make full use of the divisibility of $n - 16$ by 15; rather, we utilized a part of it, namely, the number's divisibility by 3. We see that this mathematical method is not routinely applied. Each example is a new idea.

There is, however, an essential difference between the two applications of the principle. In the problem "Knight on the Chess Board," we couldn't have reached a solution without this principle. In the present case, though, the application of the principle merely helped to find a simpler solution.

*   *   *

The problem can also be solved in a straightforward fashion. Assume that

$$\frac{n-16}{15}=x$$

and

$$\frac{n-5}{24}=y,$$

where x and y are integers.

Then we have, on the one hand, $n = 15x + 16$, and on the other, $n = 24y + 5$. Therefore, we get

$$15x + 16 = 24y + 5;$$

rearrange:

$$24y - 15x = 11,$$

and divide by 3:

$$8y - 5x = \frac{11}{3}.$$

This, however, is impossible, since there is an integer on the left-hand side, while the number on the right-hand side is not an integer.

31st Week

Dicey Question

Let a, b and c denote the values of the three dice (in the order red, blue, green). Let us follow what we obtain after each step:

$2a,$

$2a + 5,$

$5(2a + 5) = 10a + 25,$

$10a + b + 25,$

$$10(10a + b + 25) = 100a + 10b + 250,$$

$$100a + 10b + c + 250.$$

If we subtract 250 from the final result, we get a three-digit number whose digits are (in the proper order) the numbers we seek. If, for instance, the result is 484, we subtract 250 to obtain 234, and therefore, the three numbers thrown are 2, 3, and 4.

Party Question

We ask each person how many friends he/she has at the party; we add up the numbers we get in reply; call this number S.

If Mary is John's friend, then John is Mary's friend, so S is twice the number of friendships, and so S is even.

If the sum P of the odd numbers received as answers is odd, then S (which is the sum of P and all the even numbers given as answers) would also be odd, since the sum of any number of even numbers and an odd number is odd. Therefore, P is even.

If there is an odd number of answers giving an odd number, then the sum P of these numbers would also be odd, since the sum of an odd number of odd numbers is always odd. This contradicts that P is even. Therefore, an even number of answers names an odd number; that is, the number of people at the party who are friends with an odd number of people is necessarily even.

* * *

This solution is the most ingenious of the many known solutions. Its concise formulation makes the argument appear somewhat artificial. The next solution is not so ingenious, but it is quite natural.

We keep the number of people at the party fixed, and we use induction on the number of friendships at the party. (See Section 2 of the Appendix, p. 222.) If the number of friendships equals 1, that is, if at the party only two people are friends, then there are exactly two people who are friends with an odd number of people. That is, the number of people who are friends with an odd number of people is even. This verifies the induction base.

Let us assume that the statement is true if the number of friendships is k. Let the number of friendships be $k + 1$. This would come about if two people who weren't friends before became friends. If one of them had an even number of friends, and the other an odd number of friends, then their new friendship doesn't change the total number of those who have an odd number of friends (for the person with an even number of friends now has an odd number of friends, and the person with an odd number of friends now has an even number of friends). If both had either an even or an odd number of friends, the new friendship causes the number of those who have an odd number of friends either to rise or to fall by 2. The parity does not change (that is, if it was even, then it remains even; if it was odd, it remains odd), verifying the induction step.

So by induction, the statement is true.

* * *

This problem has a very interesting mathematical background. Before we get into that, let's define graphs.

A *graph* is a finite number of points on the plane; some pairs of distinct points are connected by a single line (straight or curved). The points are called the *vertices* of the graph, and the connecting lines the *edges* of the graph. (The concept defined here is called a *finite undirected graph without a loop*. In this book, we do not consider any other kind.) Note that the intersection of two edges is not necessarily a vertex. An edge has two endpoints: the two vertices it connects.

We've encountered graphs in previous puzzles. The drawings in the puzzles "Nicholas's House" (3rd Week, p. 4), "In the Garden" (11th Week, p. 15), and "Scribble" (28th Week, p. 40) are graphs.

Let's see now how we can relate this puzzle to graphs.

Represent the people at the party by points in the plane; these will be the vertices of a graph. We'll connect two vertices if the people (they represent) are friends. In this way, we obtain a graph that represents the friendships in a visual form.

This process can clearly be reversed. Every graph can be thought of as a graph that represents the friendships at a party.

The statement formulated in the puzzle can thus be reformulated in the language of graphs:

> Every graph contains an even number of vertices which are the endpoints of an odd number of edges.

The *degree* of a vertex in a graph is the number of edges of which that vertex is an endpoint. So we can rewrite the last statement:

> In every graph there is an even number of vertices of odd degree.

* * *

In the problems of *drawing without lifting the pencil*, we saw that the vertices with an odd degree played a crucial role. A figure cannot be drawn with a single, uninterrupted line if there are more than two such vertices. In the figures that can be so drawn, the number of vertices with an odd degree can be only zero or two.

First and Last Day of the Year

The number 2010 is not divisible by 4, so it isn't a leap year. So the year had 365 days.

Since $365 = 52 \cdot 7 + 1$, a year that is not a leap year has 52 weeks and an extra day. Therefore, a non-leap year ends with the same day of the week with which it begins, since a new week begins on the last day of the year.

A leap year, on the other hand, has 366 days. Therefore, in such a year, the 31st of December falls on the day of the week that follows the day of the week on which January 1st fell.

For example, 2008 was a leap year, and January 1st fell on a Tuesday. Therefore, the 31st of December fell on a Wednesday.

32nd Week

Bridge Party

Mr. North was not telling the truth.

The problem can be solved simply by listing all possible cases. We have to list all seating arrangements in which none of the four players sits in the place corresponding to his name. In all, there are nine such possible arrangements:

	North	East	South	West
(1)	West	North	East	South
(2)	West	South	North	East
(3)	West	South	East	North
(4)	East	North	West	South
(5)	East	South	West	North
(6)	East	West	North	South
(7)	South	West	North	East
(8)	South	West	East	North
(9)	South	North	West	East

Since Mr. West belonged to the winning partnership at each rubber (and since in bridge, the partnerships sit opposite one another), Mr. North won in at most two rubbers, namely, when the seating arrangement was that of (2) or (8). (The order in which the rubbers were played doesn't necessarily correspond to the listing given above, and all possible rubbers weren't necessarily played that night.)

Beside/Below Redux

In the first round, we lay out the odd cards, and the 485 even cards remain in the deck. After we've laid out card 971, card 2 goes to the bottom of the deck and we lay out card 4. Then, in the second round, we lay out the cards that are multiples of 4.

In this way, we can determine when we lay out which card. In order to explain the situation clearly, we'll speak of the first round, the second round, and so on. At the *starting position* of a round, the cards in our hand are arranged in strictly ascending order.

Thus, in the first round, card 1 is on top; in the second and third rounds, card 2 is on top; and so on. The following table summarizes the complete process of laying out the cards. The values of n are $0, 1, 2, \ldots$ and every round contains all possible n for which the sum numbers a card in the deck. So, for instance, in round 9, $150 + 512n$ is defined only for $n = 0$ and $n = 1$ since for $n = 2$ it is 1,174 and there is no such card in the deck.

Round	Number of Cards in Deck	Numbers on Cards Laid Out	Card at Bottom of Deck
1	971	$1 + 2n$	$2 + 2n$
2	485	$4 + 4n$	$2 + 4n$
3	243	$2 + 8n$	$6 + 8n$
4	121	$14 + 16n$	$6 + 16n$
5	61	$6 + 32n$	$22 + 32n$
6	30	$54 + 64n$	$22 + 64n$
7	15	$86 + 128n$	$22 + 128n$
8	8	$22 + 256n$	$150 + 256n$
9	4	$150 + 512n$	$406 + 512n$
10	2	406	918
11	1	918	

A glance at the second column shows that if there are an even number of cards, the number of cards in the deck is halved in the round. If there are an odd number of cards, the number of cards in the deck is halved and rounded up or down, in turn. In the third and fourth columns, the coefficients of n are powers of two. The numbers on the cards laid out, or the card at the bottom of the deck at the beginning of a round, are either exactly the same as the one in the previous round or increase by the sum of the two coefficients that appear in the formula for the cards laid out in the previous round.

We could reach similar general conclusions from the table, and prove them quite easily. There's no need to do this, however, since the table provides us with all the information we need.

First question: The answer is that card 918 is the last in the deck.

Second question: In which round is card 228 laid on the table? Clearly, the second round.

We can say more. The number 228 can be written in the form $4 + 4n$, since $228 = 4 + 4 \cdot 56$. The values of n started with 0, so card 228 was the 57th card laid out in the second round, and at the beginning of the round, there were 485 cards in the deck. Therefore, card 228 was laid out as the $(971 - 485) + 57 = 543$rd card.

Third question: Which card was the 634th to be laid out? In the first round, 486 cards were laid out; in the second round, $485 - 243 = 242$ cards. Since $486 + 242 = 728$ is greater than 634, the 634th card was laid on the table in the second round, indeed, as the 148th card of the second round, since $634 - 486 = 148$. This card is the one marked with the number $4 + 4 \cdot 147 = 592$.

Weight Trick III

Let the 12 boxes be represented by the 12 letters, A–L. We separate the boxes into three groups:

$$ABCD, \qquad EFGH, \qquad IJKL.$$

The following table shows the results of the weighings:

First Weighing	Second Weighing	Third Weighing	Conclusion
$ABCD > EFGH$	$ABE > CDF$	$A > B$	A heavier
		$A < B$	B heavier
		$A = B$	F lighter
	$ABE < CDF$	$C > D$	C heavier
		$C < D$	D heavier
		$C = D$	E lighter
	$ABE = CDF$	$G > H$	H lighter
		$G < H$	G lighter
$ABCD < EFGH$	$ABE > CDF$	$C > D$	D lighter
		$C < D$	C lighter
		$C = D$	E heavier
	$ABE < CDF$	$A > B$	B lighter
		$A < B$	A lighter
		$A = B$	F heavier
	$ABE = CDF$	$G > H$	G heavier
		$G < H$	H heavier

The table does not contain the third case: $ABCD = EFGH$. In this case, the box of a different weight is in the third group, $IJKL$. We consider this group, and add A, B, and C to it. Then we have four boxes, one of which is heavier or lighter than the rest, and we have three more boxes which we know have the same weight as three of the four boxes. But we've seen (in the solution of "Weight Trick I" (18th Week, p. 131)) that in such a case we can determine with two weighings which of the boxes is of a different weight, and whether it is lighter or heavier than the other boxes.

33rd Week

The Spy Who Came around the Corner

Each intersection can be reached either from the intersection above it or the one immediately to its left (when starting from the gate at the upper-left corner of the map on p. 46). So to count how many ways an intersection can be reached, add up how many ways the intersection directly to its left and how many ways the intersection immediately above it can be reached. The first intersection at the top left of the map and those crossings running along the upper wall and along the left wall can be reached only one way. Therefore, if we write on each intersection the number of ways it can be reached, we obtain a Pascal's triangle—or more precisely, a rectangular portion of one. Since the Pascal's triangle in the solution of "Carton Puzzles" (30th Week, p. 162) isn't large enough to find the 715, here is an illustration in which we reach that number. On this grid, the buildings are represented by the empty squares and the roads among them by the grid lines.

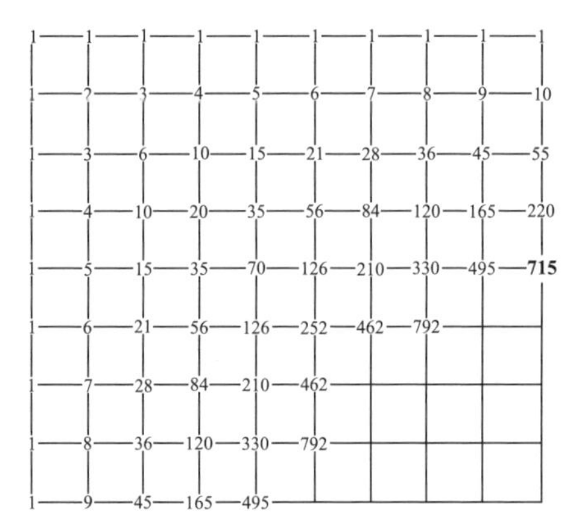

The value 715 appears at only one place. (Not all numbers are written in; the empty spots would have numbers larger than 715.) At that intersection there are two buildings, and so there are only two street corners that could be the meeting place. This is enough information to catch the spy.

Grandfather and Grandson

Clearly, the problem is solvable only if the politician was born in the twentieth century and his grandfather in the nineteenth. Otherwise, they would have to be the same age.

Let x be that two-digit number formed by the last two digits of the politician's birth year; x is the politician's age. Thus,

$$1900 + x$$

is his birth year. So we get the equation

$$1900 + x + x = 1932,$$

and we calculate that $x = 16$.

So the politician was 16 years old in 1932; he was born in 1916.

Let y be that two-digit number formed from the last two digits of the grandfather's birth year; y is the grandfather's age. Thus,

$$1800 + y$$

is the grandfather's birth year, and we have the following equation:

$$1800 + y + y = 1932.$$

From this it follows that $y = 66$.

So, the grandfather was 66 years old in 1932 and he was born in 1866.

Finally, Concerning Jealous Husbands

The crossing can be accomplished as follows:

1. Three wives cross.
2. One wife returns.
3. Two wives cross.
4. One wife returns.
5. Those three husbands whose wives are already on the far side cross.
6. One married couple returns.

7. The three husbands cross.
8. One wife returns.
9. Three wives cross.
10. One wife returns.
11. The last two wives cross.

* * *

Let n be the number of married couples, and let a_n be the smallest number so that in a boat that can carry a_n passengers, the couples can cross to the far bank (satisfying the familiar requirements).

We verified that if $n = 2$ or $n = 3$, then the problem can be solved with a two-person boat, that is, $a_n = 2$.

We've also shown that if $n > 3$, then $a_n > 2$.

Now we have seen that if $n = 5$, then $a_n = 3$.

An easy variant of the argument that a two-person boat is not sufficient to transport more than three married couples across the river can be used to show that a three-person boat won't suffice for more than five married couples.

It is, however, clear that a four-person boat can transport any number of married couples across the river. For instance, one married couple stays in the boat, and on each round trip brings another married couple across the river.

34th Week

One More Time: Knight on the Chess Board

We have to move the knight on the chessboard so that it lands on each square exactly once. We can do this in a number of ways. The illustration shows one of many solutions. The numbers written in the squares indicate the order of the moves: square 1 is the knight's starting square, the first move lands the knight on square 2, and so on.

It is really intriguing that in this diagram the sum of the numbers in each row and in each column is 260. Such an arrangement of numbers is known as a *magic square*. There are several other solutions that produce a magic square.

* * *

50	11	24	63	14	37	26	35
23	62	51	12	25	34	15	38
10	49	64	21	40	13	36	27
61	22	9	52	33	28	39	16
48	7	60	**1**	20	41	54	29
59	4	45	8	53	32	17	42
6	47	2	57	44	19	30	55
3	58	5	46	31	56	43	18

Another interesting fact: this task cannot be carried out on the smaller 4×4 chess board. It is, however, possible on the 5×5 and 6×6 chess boards. Anyone interested further in this subject should consult a book on graph theory.

Battle of Cards

Let w, x, y, and z be the money that Andy, Bert, Cris, and Dan had at the start of the game, respectively. Let's see how the financial situation of the players develops.

- After the first round:
  - Andy's money: $2w$
  - Bert's money: $2x$
  - Cris's money: $2y$
  - Dan's money: $z - w - x - y$
- After the second round:
  - Andy's money: $4w$
  - Bert's money: $4x$
  - Cris's money: $2y - 2w - 2x - (z - w - x - y) = 2y - 2w - 2x - z + w + x + y = 3y - w - x - z$
  - Dan's money: $2(z - w - x - y)$
- After the third round:
  - Andy's money: $8w$
  - Bert's money: $4x - 4w - (3y - w - x - z) - 2(z - w - x - y) = 7x - w - y - z$
  - Cris's money: $2(3y - w - x - z)$
  - Dan's money: $4(z - w - x - y)$

- After the fourth round:
  - Andy's money: $8w-(7x-w-y-z)-2(3y-w-x-z)-4(z-w-x-y) = 15w-x-y-z$
  - Bert's money: $2(7x-w-y-z)$
  - Cris's money: $4(3y-w-x-z)$
  - Dan's money: $8(z-w-x-y)$

At the conclusion of the game, each of the players has \$64, and therefore,

$$\begin{aligned} 15w-x-y-z &= 64, \\ 2(7x-w-y-z) &= 64, \\ 4(3y-w-x-z) &= 64, \\ 8(z-w-x-y) &= 64. \end{aligned}$$

The solution of this system of equations is $w = 20$, $x = 36$, $y = 68$, and $z = 132$.

Thus, at the start of the game, Andy had \$20; Bert, \$36; Chris, \$68; and Dan, \$132.

* * *

The problem can also be solved by going backwards.

After every round, the four card players together have $4 \cdot \$64 = \$256$.

In the *fourth round*, Andy loses, while the others double their money. Before the fourth round, accordingly, Bert, Chris, and Dan each had \$32. Andy, on the other hand, had $\$64 + 3 \cdot \$32 = \$160$.

In the *third round*, Bert loses, while the others double their money. Thus, before the round, Andy had \$80, Chris and Dan each had \$16, while Bert had $\$32 + \$80 + 2 \cdot \$16 = \$144$.

In the *second round*, Chris loses, and thereby doubles the others' money. Before that round, then, Andy had \$40, Bert had \$72, and Dan had \$8. Chris pays these winnings from his $\$16 + \$40 + \$72 + \$8 = \$136$.

In the *first round*, Dan loses and thereby doubles the fortunes of the others. This means that Andy originally had \$20, Bert \$36, and Chris \$68.

From this it follows that Dan, the first loser, entered the game with

$$\$8 + \$20 + \$36 + \$68 = \$132.$$

We summarize this in the following table:

	Andy	Bert	Chris	Dan
after the fourth round	64	64	64	64
before the fourth round	160	32	32	32
before the third round	80	144	16	16
before the second round	40	72	136	8
before the first round	20	36	68	132

Remains of a Long Division

Let Q be the divisor. We see that $5Q$ and $3Q$ are four-digit numbers. We have $5Q$ under the first four digits of the dividend, and the difference is a two-digit number, therefore, $5Q$ can only be a four-digit number beginning with 1. Accordingly, $1{,}000 \leq 5Q \leq 1{,}999$, and so $200 \leq Q \leq 399$.

Since $3Q$ has four digits, and it is less than $5Q$, therefore, $3Q$ also begins with 1. Additionally, the third digit of $3Q$ is 3, so $3Q \geq 1{,}030$ and so $Q \geq 344$.

Accordingly, Q cannot be greater than 399 or smaller than 344; therefore, we have a three-digit number beginning with 3. Furthermore,

$$3Q \leq 3 \cdot 399 = 1{,}197.$$

This means that $3Q$ (a four-digit number beginning with 1) has a 0 or a 1 as its second digit. Moreover, $3Q$ is divisible by 3, giving us the following possibilities:

$$1{,}032,\ 1{,}035,\ 1{,}038,\ 1{,}131,\ 1{,}134,\ 1{,}137.$$

Accordingly, the divisor, Q, is one of the following numbers:

$$344,\ 345,\ 346,\ 377,\ 378,\ 379.$$

The second subproduct is a three-digit number ending in 6. Since $3Q$ is a four-digit number, this number can only be $1 \cdot Q$ or $2 \cdot Q$. Then, the last digit of the divisor can only be 6 or 8, leaving us with only 346 or 378 from the list of possibilities above. The next to last digit of $5Q$ is 9, and five times these two numbers are

$$346 \cdot 5 = 1{,}730,$$
$$378 \cdot 5 = 1{,}890.$$

The divisor can only be 378, and the second digit of the quotient is 2.

The following illustration shows our results so far (the short lines represent digits still unknown); including the digit 5 we place at the end of the second subtraction and the digit 4 at the end of the dividend:

```
         5 2_3
378 |1_ _ _ _5 4
     1 8 9 0
       _ _ _
       7 5 6
       _ _ _5
       _ _ _ _
         1 1 3 4
         1 1 3 4
               0
```

The number above 756 can be at most 999, and so the number below 756 can be at most 2,435. The digit below the digit 5 (that we've just placed on the diagram) is 2, since $3 + 2 = 5$. Only 4 times or 9 times 378 ends in 2, and thus either 4 or 9 is the missing digit of the quotient. Choosing 9 as the missing digit leads to a contradiction: indeed, $378 \cdot 9 = 3{,}402$, and this number is greater than 2,435; the number $378 \cdot 9$ would need to be 113 less than the number above it, which cannot be greater than 2,435.

Thus, 4 is the missing digit of the quotient. Knowing the quotient and the divisor, we can now calculate the dividend and carry out the division. We obtain the following:

```
          5243
378 |1981854
     1890
       918
       756
       1625
       1512
         1134
         1134
            0
```

We can confirm that this division does, in fact, contain the required digits.

* * *

Though not emphasized at every step, it may in retrospect be stressed that every step of the solution should have been formulated as, "Since this is true, only this is possible." At the outset, we can never be sure the puzzle is actually solvable. Each individual step of this sort gives us only this certainty: if there exists a solution at all, then in this solution such and such must be fulfilled.

Of course, that's not the case with every puzzle. For example, in "Two Puzzles in Two Languages" (20th Week, p. 28, solution on p. 135), it was not necessary to confirm the correctness of our results by substitution. The equations were not only necessary, but also sufficient, conditions for the correctness of the addition; that is, the solution of the equations guarantees the correctness of the addition.

In the solution of the present puzzle, however, we found only necessary conditions. Therefore, the puzzle could only be considered solved after confirming that the conditions had been fulfilled by actually carrying out the division.

35th Week

Ten Little Slips of Paper

Among the numbers given, 16 can be represented in only two ways:

$$16 = 6 + 10,$$
$$16 = 7 + 9.$$

But if 16 is obtained from the slips with 9 and 7, then no two slips would remain that could sum to 17. Indeed, 17 can be represented in only two ways,

$$17 = 7 + 10,$$
$$17 = 8 + 9,$$

and the slips with 7 or 9 are needed in either case. So, we must have $16 = 6 + 10$ and, therefore, $17 = 8 + 9$.

Since the slips 6, 8, 9, and 10 are already used, 11 can be written only in the form $11 = 4 + 7$. This leaves us with $7 = 2 + 5$, and finally $4 = 1 + 3$.

Accordingly, Maria drew the slips with 8 and 9; Harriet, the slips with 6 and 10; Anna, the slips with 4 and 7; Louis, the slips with 2 and 5; and Greg, the slips with 1 and 3.

* * *

We can start off in another way.

From among the numbers given, 4 can only be represented in one way: $4 = 1 + 3$. Among the remaining numbers, 7 can also be written in only one form: $7 = 2 + 5$. Among the slips of paper now remaining, 11 also has a unique representation: $11 = 4 + 7$. Now, only the slips with the numbers 6, 8, 9, and 10 remain. Among these numbers, only 9 is odd and thus must be used in the representation of 17. Then, we have $17 = 8 + 9$ and, finally, $16 = 6 + 10$.

Rook Walk

Imagine that the rook can move by only one square at a time, in the permitted directions. If, for instance, the rook moves six squares in a direction, we can consider this move as composed of six individual, one-square moves. So we can reformulate the puzzle in terms of this new "one-square" rook. Clearly, the problem is solvable only if the new problem is solvable.

It's obvious that after each move the "one-square" rook lands on a square of the opposite color. Thus, we've once more reached the generalization discussed in the solution of "Knight on the Chess Board" (18th Week, p. 131), and so we see that the problem is not solvable.

* * *

This is a nice example of how it's possible, by means of this generalization, to solve a new, nontrivial problem easily and ingeniously. At the same time, we have encountered a prominent and extremely productive aspect of mathematics: generalizations can connect two widely separated problems, often letting us kill two birds with one stone.

Married Couples

Let x be the number of items bought by any one of the men, and y the number of items bought by his wife. Then for any married couple,

$$x^2 - y^2 = 63,$$

where x and y are integers and $x > y$. Since $x^2 - y^2 = (x - y)(x + y)$, it follows that

$$63 = (x - y)(x + y).$$

So, for every solution, x and y give a representation of the number 63 as the product of two integers, namely, of $x + y$ and $x - y$. Conversely, factoring the number 63 as the product of two integers, if we set the smaller of the numbers equal to $x - y$ and set the greater equal to $x + y$, we get a solution x and y.

The integer 63 can be represented as the product of two integers in three ways:

$$63 = 1 \cdot 63 = 3 \cdot 21 = 7 \cdot 9.$$

Thus, we have three systems of equations:

$$\begin{array}{llll} x - y = 1, & x + y = 63, & \text{and so} & x_1 = 32, \quad y_1 = 31; \\ x - y = 3, & x + y = 21, & \text{and so} & x_2 = 12, \quad y_2 = 9; \\ x - y = 7, & x + y = 9, & \text{and so} & x_3 = 8, \quad y_3 = 1. \end{array}$$

To each married couple belongs a pair x, y with the same index.

We know that Andy bought 23 more items than Bella, which translates into $x_i - y_j = 23$. This condition is satisfied only by $x_1 = 32$ and $y_2 = 9$. So, Andy bought 32 items, Bella bought 9, and Andy and Bella are not married.

The average price (and the number) of the items Bert bought is greater by $11 than the average value (and the number) of items bought by Anna. This condition is fulfilled only for $x_2 = 12$ and $y_3 = 1$. Bert, then, bought 12 items, and Anna bought 1. From our results so far, it follows that Bella is Bert's wife; in addition, Chris bought $x_3 = 8$ items, and his wife's name is Anna.

So Andy's wife is Crystal.

36th Week

Legal Problem from Antiquity

In its decision, the court awarded Caius 12 dinars and Sempronius 18 dinars.

The court argued as follows: Titus received five dishes, two from Caius and three from Sempronius. Therefore, the 30 dinars should be divided $2 : 3$, and that was the judgment.

At the School Dance

Let x be the number of girls.

The first girl danced with $1 + 11$ boys, the second with $2 + 11$ boys, ..., the xth with $x + 11$ boys. The number of dancing couples is thus

$$(1 + 11) + (2 + 11) + \cdots + (x + 11) = 430.$$

On the left-hand side, we have the sum of an arithmetic sequence of x terms: the difference between successive terms is a constant, in this case, 11. The sum of such a sequence is the number of terms multiplied by the arithmetic mean (average) of the first and last terms.

The first member of the sequence is 12; the last member is $x+11$. Therefore, the sum of the x members is

$$x \cdot \frac{12 + x + 11}{2} = 430.$$

Rearranging this, we get

$$x^2 + 23x - 860 = 0.$$

Using the solution formula for the quadratic equation, we get

$$x = \frac{-23 \pm \sqrt{23^2 - (4 \cdot 1 \cdot (-860))}}{2 \cdot 1},$$

which gives $x = 20$ or $x = -43$. The latter makes no sense in this puzzle, so $x = 20$.

So 20 girls attended the dance. The number of boys is equal to the number of dance partners of the last girl, that is, $20 + 11 = 31$.

You may find the two formulas used in the solution of this puzzle in Section 3 of the Appendix on p. 223.

Square Numbers

The three numbers being squared are $a - 1$, a, and $a + 1$. The number a^2 can be written as follows:

$$a^2 = \frac{(a-1)^2 + (a+1)^2 - 2}{2}.$$

The square number we seek is then

$$\frac{190{,}246{,}849 + 190{,}302{,}025 - 2}{2}.$$

The addition, subtraction, and division by 2 can be carried out with ease without a calculator or computer.

The square number we seek is thus 190,274,436.

37th Week

Half-Truths

It makes no difference how we attack the problem; we can start with the two statements of any one of the girls. Here is one approach.

Two girls assert that Rose placed fourth. If this is *not* true, then Paula (based on her statements) must have placed second and (based on Rose's statements) Kate must have placed first. But if Kate placed first, then Eve's first statement is not true; this means that it must be true that Susie placed second. However, that contradicts the statement that Paula placed second.

Thus, *Rose placed fourth*. So Paula's second statement is true, and it is not true that she placed second. Then, Kate's first statement is not true, either, and thus it is true that *Kate placed third*. Therefore, Susie's first statement is not true, and accordingly, *Eve placed fifth*. From this it follows also that Eve's first statement is not true, so *Susie placed second*. Lastly, we conclude that *Paula placed first*.

Thus, Paula placed first; Susie, second; Kate, third; Rose, fourth; and Eve, fifth. Finally, we have to check that each girl made one true statement and one untrue statement. This is true, so the problem is solved.

* * *

We can also take Kate's statements as our starting point. First, let's assume that Kate's first statement is true, namely, that Paula placed second. Then Kate did not place third, and Paula's statement that Rose placed fourth is untrue. Therefore, Rose's statement that Kate placed first is true. From this it follows that, of Eve's two statements, her claim that she placed first is untrue; therefore, it is true that Susie placed second, which again contradicts our assumption that Paula placed second.

Thus, Kate's first statement is untrue; that is, Paula did not place second. Kate's second statement, on the other hand, is true; that is, *Kate placed third*. But then Susie's first statement, that she placed third, is untrue; Susie's second statement, that *Eve placed last*, on the other hand, is true.

Now we know that Rose lied in saying that Kate won. On the other hand, Rose told the truth about herself, so *Rose placed fourth*. Then Paula's statement about Rose is true, but Paula didn't tell the truth about herself. So Paula didn't place second. If not, however, then *Paula placed first* and *Susie placed second*.

The Great Hunt

Let A be the number of points won by Andy, B by Bert, C by Chris, and D by Dan. Similarly, let b, d, w, and f be the points awarded for bagging a boar, deer, wolf, or fox, respectively.

1. At least six game animals were bagged, because at least one animal of each type was shot and at least three wolves were shot. So Dan, who bagged the most, shot at least three. Had he shot only two animals, the others could have shot at most one animal each, and in that case only five animals would have been bagged.

   Thus, $D \geq 3$.

2. The point value of the wolf is at least 2, because in the decreasing sequence, it was the next to last animal. The value of w cannot be

greater than 2. Indeed, if $w \geq 3$, then d would be at least 4 and b at least 5.

Considering that f must be at least 1, the total number of points would be greater than 18, since $3 \cdot w + f + d + b \geq 3 \cdot 3 + 1 + 4 + 5 > 18$.

Thus, $w = 2$. From this, it follows that $f = 1$.

3. Since $A + D = B + C$, and since the four of them together won 18 points, we have: $A + D = B + C = 9$.

4. The number of points D cannot be greater than 3, since Dan won the least points. If he won 4 points or more, then B and C would win at least 5, and $B + C$ would be at least 10, which contradicts that $B + C = 9$.

   Thus, $D = 3$; since $A + D = 9$, we get that $A = 6$. Either B or C is 4, and then the other is 5 (we don't know which).

5. We know from Step 1 that Dan shot three animals. Since $f = 1$ according to Step 2, therefore, *Dan killed three foxes.*

6. Chris shot the only boar, which has the highest point value, which is at least 4; he did not shoot a wolf, since, according to Step 4, $C =$ 4 or 5 and according to Step 2, $w = 2$. Similarly, we deduce that Bert bagged at most two wolves ($B \leq 5 < 3 \cdot 2$); knowing that Dan and Chris shot no wolves, it follows that *Andy bagged at least one wolf.*

7. Andy has won 6 points, of which, consistent with Step 6, two points were for a wolf. Because, aside from Dan, each hunter bagged at most two animals, Andy could only have won the other 4 points for one deer. So, *Andy killed one wolf and one deer*. Thus, $d = 4$.

8. Since $d = 4$, it follows that $b \geq 5$. Since C is at most 5 points (Step 4) and in light of Step 6, *Chris bagged just a single boar.*

9. So, consistent with Step 4, $B = 4$, and since the kill included also two more wolves, he must have shot these wolves. So, *Bert killed two wolves.*

In summary:

- The point values of *boar, deer, wolf, fox* are, in order, 5, 4, 2, 1.
- Andy bagged one deer and one wolf, winning six points ($A = 6$);
- Bert shot two wolves, winning four points ($B = 4$);
- Chris bagged one boar, winning five points ($C = 5$);
- Dan shot three foxes, winning three points ($D = 3$).

38th Week

Headscratcher with Caps

From the statement of the man sitting furthest back, we conclude that he does not see two yellow caps; otherwise, he would know there is a green cap on his head. Therefore, he sees at least one green cap.

If the man sitting in the middle would see a yellow cap, he would know that he has a green cap on his head. Since he too was unable to say what color cap was on his head, however, the man sitting in front must wear a green cap.

Assuming my friends sitting furthest back and in the middle reasoned correctly, my friend sitting furthest forward is able to declare, "There is a green cap on my head."

Schoolgirls

From the initial statement of the puzzle, it is clear that one of the ten girls is in each of the grades one through ten.

After drawing up a 10×10 table as described in the hint (p. 83), we go through the statements of the puzzle. If because of a statement a cell is to be crossed off, we enter the number of that statement in the cell (we print these numbers in bold face). If several numbers apply to a cell, we enter only one. (The small numbers will be explained later.)

Since only one of the girls is in each class, Edith cannot be in seventh grade and Frances in fifth grade, so in each of the table cells E-VII and F-V, we put a **1**.

	I	II	III	IV	V	VI	VII	VIII	IX	X
A	**2**	**2**		6	**7**	**6**	**6**	**6**	**6**	**6**
B	**5**	**5**	**5**	**5**	**5**	**6**	1	**6**	2	
C	**5**	**5**	**5**	**5**	9	4	1		2	**3**
D	**7**	**8**	**8**		**7**	4	**7**	**7**	2	**7**
E	**7**	**8**	**8**	**8**	**7**		**1**	**7**	2	**7**
F	**6**	**6**	**6**	**6**	**1**	**6**		**6**	**2**	**2**
G		5	5	5	5	4	1	5	2	3
H	**4**	**4**	**4**	6		**8**	1	9	2	3
I	**7**		7	6	**7**	**6**	**6**	**6**	**6**	**6**
J	2	2	2	2	2	2	1	2		2

In accordance with the second statement, Anna cannot be in first grade, and Frances can be neither in ninth nor tenth grade. Thus, we write a **2** in each of the cells A-I, F-IX, and F-X.

We continue this process with the next six statements; when we have finished, we see that in the row for Frances, only the column VII is empty. So we can draw the following conclusion:

Conclusion 1: Frances is in seventh grade.

Based on Conclusion 1, we know that no other girl can be in seventh grade. We indicate that by entering a small 1 in the relevant cells.

According to the second statement, Jessica is two grades above Frances, so we have the next conclusion:

Conclusion 2: Jessica is in ninth grade.

We enter a small 2 in the still empty cells that have been ruled out according to Conclusion 2.

Now, in Betty's row, only the field X is empty, and in Edith's row, only the field VI is empty. In addition, in the first column there is only one empty field, namely, in Giselle's row.

Conclusion 3: Betty is in tenth grade.

Conclusion 4: Edith is in sixth grade.

Conclusion 5: Giselle is in first grade.

We enter the small numbers 3, 4, and 5 accordingly, which leads us to the next three conclusions.

Conclusion 6: Doris is in fourth grade.

Conclusion 7: Anna is in third grade.

Conclusion 8: Irene is in second grade.

On the basis of the eighth statement, we know that Heidi is in a lower grade than Edith; therefore, given the empty cells in column V and column VIII in Heidi's row, only the cell in column V is a possibility:

Conclusion 9: Heidi is in fifth grade.

And now it is clear that

Conclusion 10: Crystal is in eighth grade.

We recognize from the solution that several statements were superfluous: for instance, parts of the seventh and eighth statements.

39th Week

Tennis Tournament

The simplest solution is that I write my report about the winner of the tournament, more precisely, about a tournament player A who won no fewer matches than anybody else (that is, nobody won more games than A). If A doesn't fulfill the condition, there would be some other tournament player B who was defeated neither by A nor by the players whom A defeated. That is, this tournament player B would have scored one victory more than A, a contradiction. Consequently, our assumption that A does not meet our condition was incorrect.

* * *

Is the winner of the tournament the only one who meets the condition? Is it possible I may write my report about other tournament players?

Before answering these questions, let's introduce a new concept: the concept of a *directed graph*.

We consider a graph, that is, a finite set of vertices of which some pairs are connected by an edge. (See the solution of "Party Question" (31st Week, p. 167).) We provide each edge with a small arrowhead to indicate that the edge not only connects the two vertices, but also has a direction. We call the set of vertices and directed edges a *directed graph*. The following illustration shows two such graphs.

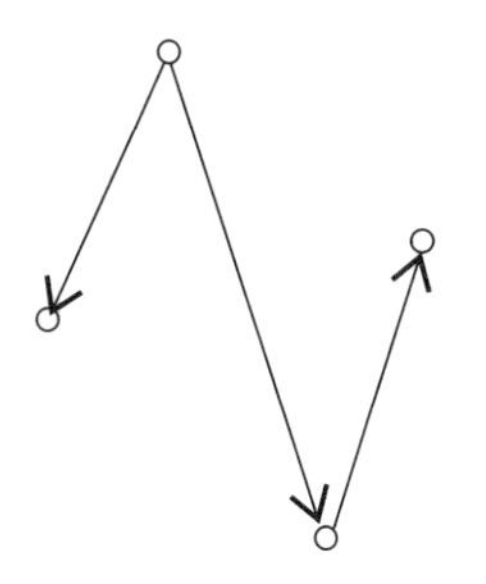

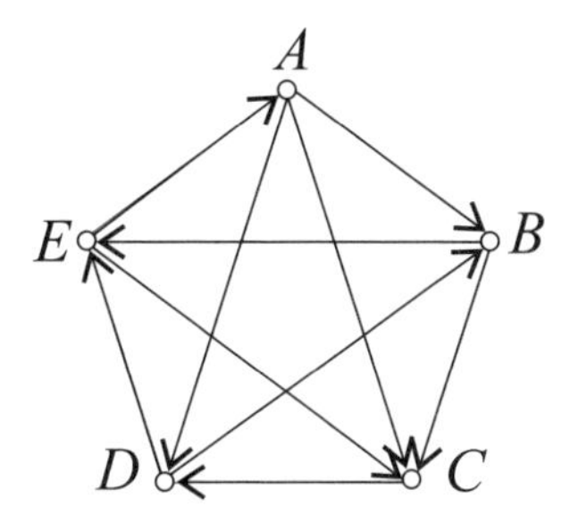

In the solution of "Party Question" (31st Week, p. 167), a graph played a prominent role in illustrating a complicated system of friendship. In the present case, a directed graph clarifies the complicated relationships of a tournament in which many players take part. We represent the players as vertices of a graph, and if A defeated B, we draw a directed edge from A to B. With such a graph, it is easy to visualize who has defeated whom.

The second of the two directed graphs above represents a tournament with five players in which each participant played every other participant. In this tournament the E meets our condition that we list all of the tournament players if we name E, all those defeated by E, and those defeated in turn by those defeated by E.

It is also possible that every player in a tournament fulfills the condition. The two graphs illustrated below show two such cases.

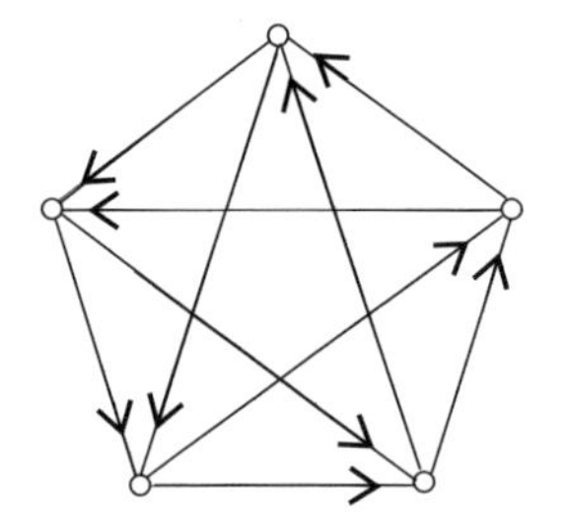

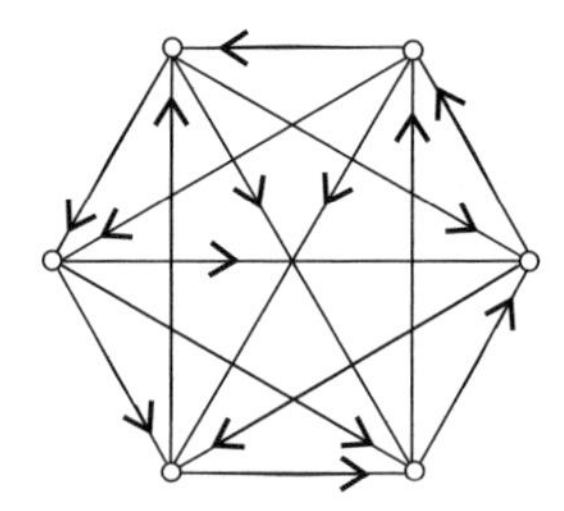

Hurdle Race

Let $5n$ be the number of \$10 gift certificates given out in the first race. We know that we can distribute the $5n$ gift certificates in five different ways. Accordingly, we have to look for integers n and k for which $5n$ can be decomposed in exactly five ways into a sum of k distinct integers, disregarding the order of the summands—that is, two sums of the same integers written in different orders will be counted as the same.

We also know that, in both contests, there were at least four runners on the winning teams because there was a runner who placed fourth. We also know that fewer than 60 gift certificates were awarded. If we go through the possibilities sequentially, we find that only two pairs of integers n and k satisfy the conditions:

- *First case:* $n = 5$ and $k = 6$. Indeed, 25 can be decomposed in exactly five ways into the sum of six distinct integers:

$$
\begin{aligned}
25 &= 1 + 2 + 3 + 4 + 5 + 10, \\
25 &= 1 + 2 + 3 + 4 + 6 + 9, \\
25 &= 1 + 2 + 3 + 4 + 7 + 8, \\
25 &= 1 + 2 + 3 + 5 + 6 + 8, \\
25 &= 1 + 2 + 4 + 5 + 6 + 7.
\end{aligned}
$$

- *Second case:* $n = 8$ and $k = 8$. Indeed, 40 can be decomposed in exactly five ways into the sum of eight distinct integers:

$$\begin{aligned}40 =&1 + 2 + 3 + 4 + 5 + 6 + 7 + 12,\\40 =&1 + 2 + 3 + 4 + 5 + 6 + 8 + 11,\\40 =&1 + 2 + 3 + 4 + 5 + 6 + 9 + 10,\\40 =&1 + 2 + 3 + 4 + 5 + 7 + 8 + 10,\\40 =&1 + 2 + 3 + 4 + 6 + 7 + 8 + 9.\end{aligned}$$

In both contests, the sum of the third- and fourth-place prizes was equal to the first-place prize. Then, for $n = 5$, only the decomposition

$$25 = 1 + 2 + 3 + 5 + 6 + 8$$

should be considered; for $n = 8$, we have the decomposition

$$40 = 1 + 2 + 3 + 4 + 5 + 6 + 8 + 11.$$

Considering now the fact that the runner who placed second on my friend's team received exactly as many gift certificates as the first-place runner on the winning team in the second contest, we see that 40 gift certificates were given out in the first contest, and each of the five teams had eight runners. In the second contest, 25 gift certificates were given out, and each team had six runners.

The question can now be answered: my friend received six gift certificates.

* * *

One might hope to come up with an elegant formula for the number of ways an integer N can be decomposed into the sum of k distinct integers, where the order of the summands is disregarded. If we can find such a formula, we would be able to solve the puzzle without trial and error.

This effort is worthwhile. Of course, such an effort won't always be successful, as it is in the present case. There is, nonetheless, a method for simplifying the trial and error by using the following mathematical theorem.

The number of decompositions of an integer N into a sum of k distinct integers (disregarding their order) is the same as the number of decompositions of the integer

$$N - \frac{k(k+1)}{2}$$

into a sum of k nonnegative integers (that is, 0 and the positive integers), where the order of the summands is disregarded.

Beware, "distinct" was dropped from the conditions.

In the puzzle, for example, we determined that the number 25 can be represented in five ways as the sum of six distinct integers (disregarding their order). Instead of this, it's also sufficient to see in how many ways

$$25 - \frac{6 \cdot 7}{2} = 4$$

can be represented as the sum of 0 and positive integers, for a total of six numbers. We can work this out easily. Here are the five possibilities:

$$\begin{aligned}4 &= 0+0+0+0+0+4,\\ 4 &= 0+0+0+0+1+3,\\ 4 &= 0+0+0+0+2+2,\\ 4 &= 0+0+0+1+1+2,\\ 4 &= 0+0+1+1+1+1.\end{aligned}$$

The theorem is easily proved, and the proof shows simultaneously how the representations of N can be obtained from the representations of $N - k(k+1)/2$. First, here is the proof.

We consider a representation of N as the sum of k distinct integers (from now on, we'll not say, "disregarding the order"):

$$x_1 + x_2 + \cdots + x_k = N \qquad (x_1 < x_2 < \cdots < x_k). \tag{1}$$

We define y_i as follows: $y_i = x_i - i$ for $i = 1, 2, \ldots k$. Then we see that the y_i are not negative, that $y_1 \le y_2 \le \cdots \le y_k$, and that

$$y_1 + y_2 + \cdots + y_k = N - (1 + 2 + \cdots + k) = N - \frac{k(k+1)}{2}. \tag{2}$$

The sum of the y_i is obtained, in other words, if we subtract the sum of the first k integers from the sum of the x_i.

In this way, we have indeed associated with every decomposition of type (1) a decomposition of type (2). Clearly, this sets up a one-to-one correspondence between the two types of decompositions.

In the following, we give side-by-side the corresponding decomposition of 4 and 25:

$$\begin{aligned}4 &= 0+0+0+0+0+4, & 25 &= 1+2+3+4+5+10,\\ 4 &= 0+0+0+0+1+3, & 25 &= 1+2+3+4+6+9,\\ 4 &= 0+0+0+0+2+2, & 25 &= 1+2+3+4+7+8,\\ 4 &= 0+0+0+1+1+2, & 25 &= 1+2+3+5+6+8,\\ 4 &= 0+0+1+1+1+1, & 25 &= 1+2+4+5+6+7.\end{aligned}$$

40th Week

Vacation Days

Taking 20 vacation days in three segments is equivalent to the problem of cutting a 20 inch measuring tape into three pieces, each of an integer length in inches (in this case, we regard different sequences of pieces of the same lengths as different).

Some possibilities are, for instance,

$$20 = 1 + 1 + 18,$$
$$20 = 1 + 18 + 1.$$

The decomposition $1 + 18 + 1$ can be related to a division of a measuring tape as follows: We cut the measuring tape at the 1 inch and 19 inch marks. In calculating the number of possibilities, then, it is sufficient to count the number of cuts that are possible.

For every cutting scheme, we choose two points at which to cut from the inch marks at 1 inch, 2 inch, ..., 19 inch. This is the same as the number of combinations of two elements chosen from 19 elements; therefore, the number of possibilities is

$$\binom{19}{2} = \frac{19 \cdot 18}{1 \cdot 2} = 171.$$

The 171 possibilities represent 171 people; they are 15% of the total personnel of the company; 10% is two thirds of 171, or 114. So, the total workforce is ten times that number, or $1{,}140$.

The company has $1{,}140$ employees.

* * *

Generally, we can formulate this problem as follows:

> In how many ways can a given integer n be decomposed into the sum of k integers provided that we regard two decompositions different if they only differ in the order of their summands?

In this puzzle, the given number equals 20 (that is, $n = 20$), and the number of the summands is 3 ($k = 3$). In the case of "The Metal Tube" (4th Week, p. 5, solution on p. 95), the given number is 85 ($n = 85$) and the number of summands is 2 ($k = 2$).

The method adopted in the solution of this problem works also in the general case. The decomposition of a given integer n into k integers in this manner can be associated with the division of an n-inch measuring tape into k pieces, each of which is a whole number of inches in length. This is the number of choices of $k-1$ elements from a set of $n-1$ elements:

$$\binom{n-1}{k-1}.$$

The Wine Grower's Estate

Each cousin received nine containers. The total amount of wine measured in quarter barrels is

$$9 \cdot 4 + 9 \cdot 3 + 9 \cdot 2 + 9 = 90.$$

Each of the cousins received a fifth of this: $90/5 = 18$ quarter barrels of wine.

Let v, w, x, y, and z denote the number of full, three-quarters full, half-full, one-quarter full, and empty containers given to one cousin. Then the following equations hold:

$$\begin{aligned} 4v + 3w + 2x + y &= 18, \\ v + w + x + y + z &= 9, \end{aligned}$$

where v, w, x, y, z are positive integers (none can be 0, since each cousin received at least one container of each kind). It is easy to see that under these conditions, there are only the following eight possibilities:

v	w	x	y	z
1	1	5	1	1
1	2	3	2	1
1	3	1	3	1
1	3	2	1	2
2	1	2	3	1
2	1	3	1	2
2	2	1	2	2
3	1	1	1	3

Each of these eight options satisfies the conditions of the will as it relates to one cousin. Additionally, we need to choose five different options for the five cousins so that we also fulfill the condition that there be exactly nine of each type of container. That is, for the five options chosen, the sums of the columns (when written in a table like the one above, but with five rows) should all equal 9.

It is clear that the option $1, 1, 5, 1, 1$ (row 1) cannot be used: If we give any one cousin five half-full containers, then we must give each of the other four one half-full container, but there are only *three* rows in which $x = 1$.

The option $2, 2, 1, 2, 2$ (row 7) can also be ruled out, because in this case $v = w$, that is, the number of full containers is the same as the number of the three-quarters-full containers, which is not permitted.

For the remaining six options, the sums of the columns (from left to right) are 10, 11, 12, 11, and 10.

We see that deleting the option $1, 2, 3, 2, 1$ (row 2) gives us the column sums we need, and the remaining five options present the solution.

The cousins could thus choose from among the following five options:

v	w	x	y	z
1	3	1	3	1
1	3	2	1	2
2	1	2	3	1
2	1	3	1	2
3	1	1	1	3

Andy received the most empty containers; for him, therefore, $3, 1, 1, 1, 3$ (row 5 in the new table) is the match. Bert and Dan received the most quarter-full containers; for them, therefore, the options are $1, 3, 1, 3, 1$ (row 1) and $2, 1, 2, 3, 1$ (row 3). Dan and Eddy received the fewest full containers, so for them, the options are $1, 3, 1, 3, 1$ (row 1) and $1, 3, 2, 1, 2$ (row 2).

Therefore, Dan's barrels match $1, 3, 1, 3, 1$ (row 1); for Bert, the match is $2, 1, 2, 3, 1$ (row 3), for Eddy, $1, 3, 2, 1, 2$ (row 2), and for Chris, $2, 1, 3, 1, 2$ (row 4).

The following table shows how the cousins divided the inheritance among themselves:

	Andy	Bert	Chris	Dan	Eddy
full	3	2	2	1	1
3/4-full	1	1	1	3	3
1/2-full	1	2	3	1	2
1/4-full	1	3	1	3	1
empty	3	1	2	1	2

41st Week

Divine Injunction

This problem is easier to solve by first generalizing it. Let's examine the more general question: what is the minimum number of steps required to transfer n disks from the first needle to another needle? Let a_n be this minimal number of steps. We will show that $a_n = 2^n - 1$ and obtain then the solution of the puzzle by setting $n = 64$.

We prove this statement by induction.

Our formula is correct for $n = 1$, since one disk can be transferred in one step from the needle holding the disk onto another needle, and thus

$$a_1 = 2^1 - 1 = 1$$

is, in fact, correct. (Induction base, see Section 2 of the Appendix, p. 222.)

Let's assume, now, that the statement is already proved for $n - 1$: that is,

$$a_{n-1} = 2^{n-1} - 1.$$

Using the statement for $n-1$, we have to show that the statement is also true for n, that is, that the correctness of the statement remains valid as we pass from $n - 1$ to n. For this, we concentrate—in the process of transfer by the minimal number of steps—on the move when the disk having the greatest diameter (originally lowermost) is transferred to another needle. This step can be carried out only when the other $n - 1$ disks have all been moved onto a third needle.

Since the last disk finally to be transferred has the largest diameter, there is no disk below it. But there is no disk above it either or we couldn't move it. There is no disk on the needle onto which we intend to transfer all the disks, for this disk may not be placed on top of any disk of smaller diameter. This position can be reached by starting at the initial position and transferring the upper $n - 1$ disks to the third needle.

By the induction hypothesis,

$$a_{n-1} = 2^{n-1} - 1$$

is the smallest number of steps required to carry this out. After those steps, the nth (largest and lowest) disk can be transferred to the second needle. Then, another $n - 1$ disks have to be transferred from the third needle to the second one, where the disk of largest diameter was placed. In these transfers, the nth disk plays no role, since it will not be moved from its present location, and since it has the largest diameter, any other disk may freely be laid on top of it.

According to the induction hypothesis, $a_{n-1} = 2^{n-1} - 1$, the minimal number of steps required for the second transfer of the $n - 1$ disks is likewise

$$a_{n-1} = 2^{n-1} - 1.$$

We obtain the number of steps necessary for the transfer of n disks as follows:

$$a_n = a_{n-1} + 1 + a_{n-1} = 2^{n-1} - 1 + 1 + 2^{n-1} - 1 = 2 \cdot 2^{n-1} - 1 = 2^n - 1.$$

Based on the assumption that the assertion is correct for $n-1$, we were able to conclude that it is also correct for n. And because we saw that the assertion is correct for $n = 1$, the correctness passes on from 1 to 2, from 2 to 3, and so on. In other words, the assertion is correct for all integers n.

In the puzzle, $n = 64$, and therefore, $2^{64} - 1$ steps are required to achieve the transfer.

This number is approximately $1.845 \cdot 10^{19}$, and that many seconds make up more than 500 billion years.[8] So even if the divine injunction was pronounced a few thousand years ago, we don't have to worry about an imminent end of the world.

* * *

From the proof above it follows also that the problem has a solution. At the same time, the proof describes a method for carrying out the transfer. In addition, we realize there is a single shortest solution, for the method described for carrying out the transfer is not only sufficient, but also necessary.

Mathematicians in Conversation

Let's look first at Andy's (and Chris's) age. Let the base of the number system be q, and let x be the first of the four one-digit numbers. Since we are not dealing here with the decimal system, q is an integer not equal to 10. Further, x is a digit in the base-q number system, and therefore, x is a nonnegative integer, not greater than $q - 1$. Andy's statement can be represented as follows (where x, $x+2$, $x+4$, and $x+6$ are the four numbers he references that succeed one another by a difference of two each):

$$qx + x + 2 = (x + 4)(x + 6). \tag{1}$$

Rearrangement gives

$$x(q + 1 - x - 10) = 22. \tag{2}$$

In light of (2), we see that x is a positive factor of 22. We consider the possibilities in order:

- If $x = 1$, then $q + 1 - x - 10 = 22$, and it follows that $q = 32$.
- If $x = 2$, then $q + 1 - x - 10 = 11$, and it follows that $q = 22$.

[8] This number precisely is 18,446,744,079,551,615.

- If $x = 11$ or $x = 22$, then by (1), we would obtain a number too large (greater than 200) to be the correct age.

So only $x = 1$ or $x = 2$ will give us a solution.

The solution $x = 1$, $q = 32$ yields for the age

$$32 \cdot 1 + (1 + 2) = (1 + 4)(1 + 6) = 35 \text{ years.}$$

The solution $x = 2$, $q = 22$ yields for the age

$$22 \cdot 2 + (2 + 2) = (2 + 4)(2 + 6) = 48 \text{ years.}$$

Thus, Chris, the younger, is 35 years old, and Andy, the elder, is 48 years old (35 in base-32 is 13; 48 in base-22 is 24).

We'll now write out what Bert said in a base-t system:

$$tz + z + 1 = (z + 2)(z + 3);$$

rearranging this gives

$$z(t + 1 - 5 - z) = 5. \qquad (3)$$

Accordingly, z is a positive factor of 5, that is, $z = 1$ or $z = 5$.

If $z = 1$, then by (3), we get that $t = 10$, and the age is

$$3 \cdot 4 = 12 \text{ years,}$$

which isn't possible, since the 35-year-old Chris is the youngest.

If $z = 5$, then again by (3) we get that $t = 10$, and the age is

$$7 \cdot 8 = 56 \text{ years.}$$

Accordingly, Bert is 56 years old, and we see also that Bert is absent-minded, for the property he stated holds only for a number in the decimal system; recall that at the start, they had agreed they would give their ages in a number system other than the decimal.

42nd Week

Gifts

Eighty dollars must be divided up as the sum of four parts such that each part is a positive, integer multiple of ten (here, we do not recognize two divisions as differing from one another if they differ only in the sequence of the terms).

We get five different possibilities:

$$80 = 10 + 10 + 10 + 50, \quad (1)$$
$$80 = 10 + 10 + 20 + 40, \quad (2)$$
$$80 = 10 + 10 + 30 + 30, \quad (3)$$
$$80 = 10 + 20 + 20 + 30, \quad (4)$$
$$80 = 20 + 20 + 20 + 20. \quad (5)$$

Each girl had to choose a different one. (If Mr. Smith had more than five daughters, he would have had to plan out the problem posed to his daughters differently.)

Frances spent more for Bert than for the other three combined. That could have happened only if she chose option (1).

Mary and Eve chose a solution in which there is a single greatest number. Of the options remaining, only (2) and (4) satisfy this condition. Accordingly, Mary and Eve chose either option (2) or option (4).

Clara spent on Simon and Robert combined as much as Frances spent for the other two boys, Bert and Josh. We know that Frances spent $50 for Bert and $10 each for Josh, Robert, and Simon. So, Clara spent $60 for Simon and Robert combined. With option (5), that can't happen, and therefore, Clara could have shopped only by option (3). Accordingly, she spent $30 each on Simon and Robert, and $10 each on Bert and Josh.

Now we know, then, that Susie has shopped by option (5), and thus for each boy, she spent $20.

The following partial solution table shows what we know up to this point:

	Amount spent on			
	Bert	Josh	Robert	Simon
Frances	50	10	10	10
Clara	10	10	30	30
Susie	20	20	20	20

We know too that the five girls combined have spent an equal amount on each boy. Since they have spent a total of $5 \cdot \$80$ or $400, a total of $100 was allotted to each boy.

In Eve's and Mary's cases, we need to consider this condition. From the partial solution table, Josh needs $60 to come to $100. From options (2) and (4), this is possible only by using the amounts $20 + $40. Mary spent the most on Josh, and Eve the most on Robert. Therefore, option (2) fits Mary's case, and option (4) fits Eve's.

So now in Robert's column (in Eve's row) we put the largest number from option (4), namely, 30; then we can determine the missing amounts, given the condition that every boy receives gifts of a total value of $100.

The solution is given in the following table:

	Amount spent on			
	Andy	Josh	Robert	Simon
Frances	50	10	10	10
Clara	10	10	30	30
Susie	20	20	20	20
Mary	10	40	10	20
Eve	10	20	30	20

The Fickle Sultan

In the end, those cells get unlocked where the guards have turned the lock an odd number of times. On the first pass, the guard turns every lock once, and he is opening every lock. If a lock is turned a second time, it is locked again. The third time, it is opened, and so on.

If we number the cells consecutively, starting with the first door, from cell 1 to cell 100, we can reformulate the previous statement:

> At the end of the process, those cells will be open the numbers of which have an odd number of positive integers as factors.

The question is, then, which of the first 100 integers have an odd number of positive integers as factors (count as factors 1 and the number itself also).

The factors of the numbers can be grouped into *factor pairs*. If, for example, a number n is divisible by 6, then n is also divisible by $n/6$; the numbers 6 and $n/6$ form factor pairs if $6 \neq n/6$. Consequently, a number has an odd number of factors only if some factor does not belong to a pair. So we see that the squares of integers have an odd number of integer factors, and that this cannot be the case with other integers.

As an example, we'll arrange the factors of $6^2 = 36$ in pairs: 1 and 36, 2 and 18, 3 and 12, 4 and 9, and finally 6, which belongs to no pair.

Thus, a total of ten prisoners go free, namely, those occupying cells 1, 4, 9, 16, 25, 36, 49, 64, 81, and 100.

43rd Week

Military Band—Blow Your Own Horn!

Since no officer's wife could be the sister of the husband of her husband's sister, it is clear that each officer has two brothers-in-law: his sister's husband and his wife's brother. We can thus arrange the officers in a cyclical graph so that on either side of every officer are his two brothers-in-law.

Let A, B, C, D, and E be the five officers, and let

$$A - B - C - D - E - A$$

be the cyclical graph where edges – connect brothers-in-law.

Davis's two brothers-in-law were on tour with the band in France, along with Patterson's two brothers-in-law. We can't say the same, though, for either of the colonel's brothers-in-law. If B is the colonel, then A and C weren't in France. Then, B, D, and E must have gone to France, and A and C must be Davis and Patterson. So we can assume that *A is Davis* and *C is Patterson.*

Further, B, C, and D were in Finland. The lieutenant is thus either A or E, and likewise Vance is A or E.

Either C and D or C and B were not in Canada. Since B is the colonel, we know he wasn't in Canada. From this it follows that A, D, and E were in Canada. So E is the lieutenant colonel. Consequently, A is the lieutenant.

We know, too, that A, B, and C were together in Japan. So E and D weren't in Japan, but Vance also was not there, and Vance is A or E. Thus, *Vance is E.*

We know by now that E is the lieutenant colonel, and that the captain too was not in Japan. Thus, D is the captain. And, accordingly, C is the major.

Matthews can now be only B or D. But at least one of Matthews's brothers-in-law holds a higher rank than Matthews himself. So, *Matthews is D* (the captain). Finally, *Burnes is B* (the colonel).

In summary,

Name	Rank	Wife's Maiden Name
Davis	Lieutenant	Burnes or Vance
Burnes	Colonel	Davis or Patterson
Patterson	Major	Burnes or Matthews
Matthews	Captain	Patterson or Vance
Vance	Lieutenant Colonel	Davis or Matthews

At the Round Table

Let's designate the married couples with the letter pairs: Aa, Bb, Cc, and Dd. Since we are dealing with a round table, without loss of generality, we can assume that A always sits at the same place.

We'll now determine the number of those seating arrangements in which at least one couple sits together (which we'll have to subtract from the number of possible seating arrangements at the round table):

(a) How many options are there in which all four couples sit together? If Aa's place remains fixed, then for the seating of the other three couples, there are $3!$ options.

If, for each option, we switch the seating of the spouses, then we obtain $(2!)^4 = 16$ seating options. The number of seating options is thus

$$3! \cdot 16 = 6 \cdot 16 = 96. \qquad (1)$$

(b) How many options are there in which the three couples sit together, but the fourth couple is separated? Let's assume the couples Aa, Bb, and Cc sit together. Then we can arrange Aa, Bb, Cc, D, and d (with the places of Aa fixed) in $4!$ different ways.

If we interchange the places of the couples sitting together, we get $2^3 = 8$ options.

Thus, we have a total of $4! \cdot 8 = 24 \cdot 8 = 192$ options. But this number includes the seating arrangements already counted in (a), and so we get

$$192 - 96 = 96. \qquad (2)$$

For the selection of three couples from four couples, there are

$$\binom{4}{3} = \binom{4}{1} = 4$$

possibilities. (The number of ways to choose three elements from a set of four is determined most easily by noting that when three elements have been chosen, a single element remains. Knowing this element not chosen determines the three chosen ones.)

The number of seating options in which exactly three couples sit together is thus

$$96 \cdot 4 = 384.$$

(c) How many options are there in which two couples sit together, while the other two couples are separated? Let's assume that couples Aa and Bb are seated together. Then, we can arrange Aa, Bb, C, c, D, and d (the location of Aa remains fixed) in $5!$ ways.

Interchanging the places of the couples seated together, yields $2^2 = 4$ options. So, we have a total of

$$5! \cdot 4 = 120 \cdot 4 = 480$$

options. But these, just as in (b), include the seating arrangements in (a), whose number is given in (1). Also, the number calculated in

(2) is, in fact, contained twice in this number (once, when the couple Cc are the third couple to sit together, and again when as the third couple Dd sit together). Therefore, we need to subtract these numbers:

$$480 - 96 - 2 \cdot 96 = 480 - 288 = 192. \tag{3}$$

There are

$$\binom{4}{2} = 6$$

ways we can choose two couples out of four. The number of seating options in which exactly two couples sit together is thus

$$192 \cdot 6 = 1{,}152.$$

(d) Last, we'll look at the option: just one couple sits together, and the other couples sit separately. We'll assume couple Aa sits together. Then, we can arrange Aa, B, b, C, c, D, and d (while Aa's seating location remains fixed), in $6!$ different ways. For each option, interchanging the places of the one couple seated together, A and a, yields two options. So, we have a total of

$$6! \cdot 2 = 720 \cdot 2 = 1{,}440$$

options. This number, however, includes the seating arrangements figured out in (1) once, and the seating arrangements given in (2) are counted three times (for the spouses seated together, we can choose two of the other three married couples in three different ways). Furthermore, from (c), those seating arrangements whose number is given in (3) are included and have been counted three times (since for the spouses seated together, we can choose one of the other three couples in three different ways). Therefore, we still need to subtract the number of these seating arrangements:

$$1{,}440 - 96 - 3 \cdot 96 - 3 \cdot 192 = 1{,}440 - 96 - 288 - 576 = 480.$$

There are four ways of choosing one couple out of four; the number of seating arrangements in which the spouses of exactly one couple are seated next to each other is thus

$$480 \cdot 4 = 1{,}920.$$

We obtain the number of seating arrangements we seek by subtracting the sum of the seating arrangements computed in (a), (b), (c), and (d) from the total number of possible seating arrangements.

In how many ways can eight people seat themselves at a round table? On a bench, there are $8!$ different sequences in which eight people can arrange themselves. But if we seat people at a round table, we have a different situation. If we move each person seated in a given order to the place of their immediate neighbor to the right, this arrangement can't be distinguished from the original seating arrangement, since a round table has no singular starting or ending point. We can repeat this shift by one place eight times until we have returned to the starting position. Accordingly, we can reduce the $8!$ cases into groups of eight such that we cannot distinguish the seating arrangements from one another. Thus, for our table there are

$$\frac{8!}{8} = 7!$$

different seating arrangements. (In the previous computations, we took the roundness of the table into account by maintaining one fixed place.)

For the total number of solutions we seek, we finally obtain

$$7! - (96 + 384 + 1{,}152 + 1{,}920) = 5{,}040 - 3{,}552 = 1{,}488.$$

We see here that the person who thought there were enough different seating arrangements to last several weeks was entirely correct. In fact, there are new seating arrangements to last several years, since generally the couples eat lunch at the restaurant five times a week: a year has 52 weeks, and $5 \cdot 52 = 260$, and even five times this number is smaller than $1{,}488$.

* * *

Our starting point that a round table has no distinguished seat is of course a mathematical abstraction. In reality, it might well be that one seat at the table is preferable, because, let's say, it is nearer to the exit, or because it offers a better view of the room.

44th Week

WaterMonster

At first glance, it seems the solution depends on whether the year is a leap year. But if we give it a bit more thought, we recognize—just as in the solution of "The Fickle Sultan" (42nd Week, p. 198)—that water flows from a nozzle on December 31, if the integer corresponding to the nozzle has an odd number of factors.

Let's look at what happens with the 365th nozzle. Whether or not it is a leap year (because 365 has the four factors 1, 5, 73, and 365), the automatic control opens the 365th nozzle at the start of the year (on January 1st, at midnight), shuts it off at the start of the fifth day (on January 5th), opens it again at the start of the seventy-third day, shuts it off at the start of the 365th day. On December 31st, the valve is closed (whether or not it is the 365th or the 366th day of the year).

A valve is opened on December 31st, then, if its number has an odd number of factors. But this can occur only if the number is the square of an integer (see "The Fickle Sultan," 42nd Week, solution on p. 198).

Thus, on December 31st, water flows from 19 nozzles, namely, those numbered 1, 4, 9, 16, 25, 36, 49, 64, 81, 100, 121, 144, 169, 196, 225, 256, 289, 324, and 361.

Tommy's Surprise

Let us rewrite the multiplication in the form

$$abTcdeT \cdot fghiTj,$$

using the letters a–j for the missing digits. Note that different letters may stand for the same digits.

Since the six subproducts are different, it follows that f, g, h, i, T, and j are different from each other and 0 (only T could be 0, but then there would be a subproduct consisting only of Ts).

Certainly, T cannot be 1, 5, and 6, because the last digit of the second subproduct (which we get by multiplying with T) shows that $T \cdot T$ does end with T.

The fourth subproduct (which we get by multiplying with h) differs from the multiplicand, so $h \neq 1$, and ends with T, which is possible only if $h = 6$ and $T = 2$, 4, or 8 (here we use the statement we proved in the hint on p. 85, and also that $h \neq 1$, $T \neq 5$, and $T \neq 6$).

We know that $T \neq 2$, because in the fourth column from the left, the sum

$$T + T + T + y$$

ends with T, where y (as a number carried from the fifth column) can in no case be greater than 5.

Also, $T \neq 8$, since in case $T = 8$, the eight-digit fourth subproduct, obtained from the multiplication of the seven-digit multiplicand by 6, can clearly not begin with 88. Thus, $T = 4$ is the only possibility.

The third subproduct (which we get by multiplying with i) thus ends with 4, that is, the product $i \cdot 4$ ends with 4. Since the digits of the multiplier are different and $h = 6$, only $i = 1$ is possible.

The fourth subproduct (which we get by multiplying with 6) begins with 44, and thus a can only be 7. Multiplying by b, we carry 2, which is possible only in case $b = 3$ or $b = 4$; but 4 is eliminated, since $b \neq T$. So $b = 3$.

In the product, the first digit from the left is 4. If we consider what numbers may be carried from the previous column and consider that the first digit of the last subproduct cannot be 4, then that first digit can only be 3. Therefore, f can only be 5.

Therefore, the last subproduct ends with 0, and its fourth digit can only be 4 if the number to be carried from $5c$ is 4. So $c = 8$ or $c = 9$. But $c = 8$ is not possible, since the sum of $6c$ and the number carried in the fourth subproduct, obtained from the previous product, ends in 4. This is possible only if the number carried from $6d$ equals 6, and that in turn would require that $e = 9$, and the number carried from multiplying 6 times 4 equals 6. This is impossible, however, and therefore, $c = 9$.

From the fifth subproduct it follows that the sum of $9g$ and the number carried from the previous product, the sum of $4g$ and the previous number carried, as well as the sum of $3g$ and the previous number carried, all end with 4. If we consider in order all the possibilities, we see that only $g = 7$ fulfills the conditions.

In the fifth subproduct, $7e + 2$ ends with 4, and thus $e = 6$.

If we likewise multiply 9 by 7 in the fifth subproduct, then 4 is the last digit. And since 7 times 9 is 63, we obtain 1 as the number carried from $7d + 4$, and so $d = 1$.

For j, only 2, 3, 8, and 9 remain as possibilities. Taking into consideration that 734 are the first three digits of the multiplicand, we see that only $j = 2$ can produce the first subproduct.

And here is the product we found:

```
         7349164
      ×   576142
   -------------
        14698328
       29396656
       7349164
     44094984
    51444148
   36745820
   -------------
   4234162045288
```

45th Week

A Number in Mind

Let x be the number Alfred has in mind. Interchanging the ones digit with the tens digit of its square yields the square of $x + 1$.

If the last digit of a number is

$$0, 1, 2, 3, 4, 5, 6, 7, 8, 9,$$

then the last digit of its square is,

$$0, 1, 4, 9, 6, 5, 6, 9, 4, 1,$$

respectively. Since the last digit of $(x + 1)^2$ equals the tens digit of x^2, the last two digits of x^2 must be one of the following:

$$10, 41, 94, 69, 56, 65, 96, 49, 14, 01.$$

The numbers with the last two digits 10, 94, and 14 are even, but not divisible by 4; however, an even square number must be divisible by 4 (see the hint on p. 86). So, these are ruled out.

Similarly, 65 must also be ruled out, since a number ending in 65, though divisible by 5, is not divisible by 25, and thus cannot be a square number. (This is analogous to the fact that the square of an even number is also divisible by 4.)

We have eliminated 10, 94, 14, and 65. Interchanging the digits, we get 01, 49, 41, and 56. These can also be ruled out because if x^2 ends with 01, 49, 41, and 56, then $(x + 1)^2$ would end with 10, 94, 14, or 65, which is not possible. Accordingly, the final two digits of the number we're looking for could be only 69 or 96.

Since the leading digits of the two square numbers agree, their difference is

$$96 - 69 = 27$$

or

$$69 - 96 = -27.$$

Since

$$(x + 1)^2 - x^2 = 2x + 1,$$

there are two possibilities:

$$2x + 1 = 27,$$
$$2x + 1 = -27.$$

From this it follows that $x = 13$ or $x = -14$. Since Alfred had a positive integer in mind, it is 13.

In fact, reversing the last two digits of $13^2 = 169$, we get $14^2 = 196$.

The Hungry Sales Woman

Let x, y, and z be the numbers of apples in the three baskets, respectively. We are looking for the integer solutions of the equation

$$x^2 + y^2 + z^2 = x^2y^2.$$

The equation can be rewritten as

$$1 + z^2 = x^2y^2 - x^2 - y^2 + 1 = (x^2 - 1)(y^2 - 1).$$

We claim that the two sides of the equation are odd. Let's assume that the left-hand side is even. Then z^2 is odd, and so is z. The square of an odd number is of the form $4k + 1$, for an integer k, so, the left-hand side would have the form $4k + 2$ (see the hint to the previous puzzle, "A Number in Mind," on p. 86).

The right-hand side, on the other hand, can be even only if at least one of its factors is even, say, $x^2 - 1$. If this is the case, then $x^2 - 1$ is of the form $4k$, for some integer k (since x^2 is odd and therefore, has the form $4k + 1$). This means that the right-hand side would be divisible by 4, but not the left-hand side, which is impossible.

Therefore, the two sides of our rewritten equation must be odd, and x, y, and z are even. Let us write x, y, and z in the form: $x = 2^a t$, $y = 2^b u$, and $z = 2^c v$, where a, b, c, t, u, v are integers and t, u, and v are odd. Note that $a \geq 1$, $b \geq 1$, and $c \geq 1$.

Our original equation, then, becomes

$$2^{2a}t^2 + 2^{2b}u^2 + 2^{2c}v^2 = 2^{2a+2b}t^2u^2.$$

By reason of symmetry, we can assume that a is the smallest of the integers a, b, and c. We divide our equation by the 2^{2a} and obtain

$$t^2 + 2^{2(b-a)}u^2 + 2^{2(c-a)}v^2 = 2^{2b}t^2u^2.$$

Since t^2 is odd, the left-hand side can be even only if one of the other two summands is even and the other is odd. That is, of the two differences, $b - a$ and $c - a$, one is 0, while the other isn't. In this case, however, the left-hand side is the sum of the squares of two odd numbers and of one even number. Since the two odd numbers have the form $4k + 1$, and the even number has the form $4k$, the left-hand side takes the form $4k + 2$, while the right-hand side is divisible by 4.

This contradiction is resolved only if u, t, and v are all 0, and therefore, x, y, and z are also 0. (The contradiction is not even resolved if $a = b = c = 0$, but one of the numbers t, u, or v is different from 0, because then the left-hand side has the form $4k+3$, while the right-hand side has the form $4k+1$).

So, the market woman went hungry because there were no apples in any of her baskets.

46th Week

Another Fragmentary Division

Observation 1: Seven times the three-digit divisor is a three-digit number with the following property: if this number is subtracted from a three-digit number, the difference is also a three-digit number.

Observation 2: The product of the third digit of the quotient and the divisor is also a three-digit number, which is larger than seven times the divisor, since subtracting it from a four-digit number yields only a two-digit difference.

Observation 3: The first and last digits of the quotient are larger than the third digit, since when multiplied by the divisor, they both yield a four-digit number.

From these observations it follows that the third digit of the quotient must be 8, while the first and last digits must both be 9.

Observation 4: Two digits of the dividend had to be carried down for the last subproduct, and therefore, the fourth digit of the quotient is 0. The quotient is, thus, 97,809.

Observation 5: If x is the three-digit divisor, then $8x < 1{,}000$. So $x < 125$.

Based on Observations 4 and 5, we observe: subtracting 8 times the divisor from a four-digit number yields a two-digit number with the property that, by adding a third digit, the resulting three-digit number is smaller still than the divisor, which is itself smaller than 125. Therefore, we have

$$1{,}000 - 8x \leq 12,$$

and so

$$x \geq \frac{988}{8} = 123.5.$$

From this, in accordance with Observation 5, we get the number 124 as the divisor.

Thus, the dividend is

$$97{,}809 \cdot 124 = 12{,}128{,}316.$$

Now we carry out this division to see that it does, in fact, satisfy the conditions of the puzzle. Thus, the original division looked like this:

```
          9 7 8 0 9
        -----------------
1 2 4 | 1 2 1 2 8 3 1 6
        1 1 1 6
        -------
            9 6 8
            8 6 8
            -----
            1 0 0 3
              9 9 2
            -------
                1 1 1 6
                1 1 1 6
                -------
                      0
```

Return of the Sultan

A prisoner is freed from his cell if the number of turns of the cell lock is odd.

Just as in "The Fickle Sultan" (42nd Week, p. 60, hint p. 84, solution p. 198), a cell's door is visited by a guard as many times as the number of divisors of its cell number. This time, though, the guard gives not just one turn to the lock, but the number of turns indicated by the divisor.

We now see readily that to find out which cells stand open at the conclusion of the process, we need to find those integers, not greater than 100, with the property that the sum of its divisors is odd.

The sum of the divisors can be odd only if the number of odd factors is odd. Each number can be written in the form $2^n \cdot k$, where n is a nonnegative integer and k is an odd integer (in the cases of odd numbers, $n = 0$), and the odd divisors of a number are precisely the divisors of k. The number of divisors of k is odd only if k is a square number (here, we again use the argument developed in the solution of "The Fickle Sultan" (p. 198)).

Accordingly, the numbers we seek have the form

$$a^2 \cdot 2^n,$$

where a is an odd integer, and n is a nonnegative integer. So, the occupants of cells numbered

$$1,\ 2,\ 4,\ 8,\ 9,\ 16,\ 18,\ 25,\ 32,\ 36,\ 49,\ 50,\ 64,\ 72,\ 81,\ 98, \text{ and } 100$$

are set free.

* * *

Note that the numbers sought can also be described as follows. They are integers of the form

$$2^n \quad (n \geq 0)$$

and of the form

$$2^n \cdot (2k+1)^2 \quad (n \geq 0, \quad k \geq 1).$$

47th Week

An Interesting Game

The starting player always wins by pursuing the following strategy. He lays the first quarter in the exact center of the rectangular table. Whenever his opponent lays a coin on the table, the first player responds by placing his coin symmetrically opposite it (with respect to the midpoint of the table). The beginning player will clearly lay the last coin on the table, since the spot lying symmetrically to the coin last placed by his opponent will be open.

* * *

We must note that we are dealing with a theoretical game, hard to realize in practice. It's simply not possible to position a coin precisely in the middle of the table or, for that matter, in an exactly symmetrical position.

At Lake Michigan V

In the puzzles "At Lake Michigan" (23rd Week, p. 33; 24th Week, p. 35; 27th Week, p. 39; and 29th Week, p. 42), we presented the game called *Nim*.

In these puzzles and solutions, we described various winning strategies. Now we shall verify that they actually work.

If there is at most *one* pile with more than one match, then the person who makes the next move clearly wins.

So let us assume that there is more than one pile and at least two piles have more than one match. Let Steve have a Binary Winning Position (in particular, it is Tommy's turn). We want to verify that Steve can win.

It is Tommy's turn, so he removes some matches from a pile, say, 1,001 matches (in binary form, that is, 9 matches). That will make the first and fourth column from the right odd. Steve's job is to remove matches from a pile to make the columns even again.

Since the fourth column became odd, there must be another pile with at least 2^3 matches. Steve removes $2^3 - 1$ matches from this pile, thereby decreasing by 1 the fourth column and increasing by 1 the first column, making both even again.

Similarly, whatever Tommy removes can be countered by Steve to make the columns even again.

A formal proof is almost the same as this example, except that we have to use the general inequality

$$1 + 2 + 2^2 + \cdots + 2^{n-1} < 2^n.$$

Let's look at one more example of the application of the strategy.

It is Steve's turn, and the matches are in three piles: 13 in one pile, 10 in another, and 29 in the last. Steve writes these numbers in the binary system and adds up the numbers of the corresponding powers of two:

	16's	8's	4's	2's	1's
13		1	1	0	1
10		1	0	1	0
29	1	1	1	0	1
Sum:	1	3	2	1	2

The 16's, the 8's, and the 2's occur an odd number of times. In the case of 29, a 16's occurs, and thus for this pile, we remove a total of $16 + 8 - 2$, that is, 22 matches (the 2 has a minus sign because 29 has a 0 in the 2's place). Accordingly, 7 matches remain in the third pile, and thereby every power of 2 now does occur an even number of times:

	16's	8's	4's	2's	1's
13		1	1	0	1
10		1	0	1	0
7			1	1	1
Sum:		2	2	2	2

* * *

The Nim game has many variations. It is, for instance, also played with the rule that the person removing the last match wins. The game can be found in many variant forms on the Internet.

48th Week

Computational Wizard

First, we'll reduce the numbers into prime factors and interchange the two denominators:

$$\frac{2 \cdot 3 \cdot \left(3^3\right)^{12} + 2 \cdot \left(3^4\right)^9}{\left(3^2\right)^{19} - \left(3^6\right)^6} \cdot \frac{2^4 \cdot 5 \cdot \left(2^5\right)^3 \cdot \left(5^3\right)^4}{\left(2^3 \cdot 2^6 \cdot 5^6\right)^2}.$$

By simplification, we effortlessly obtain the value of the complicated looking number:

$$\frac{2 \cdot 3^{37} + 2 \cdot 3^{36}}{3^{38} - 3^{36}} \cdot \frac{2^{19} \cdot 5^{13}}{2^{18} \cdot 5^{12}} = \frac{2 \cdot 3^{36}(3+1)}{3^{36}\left(3^2 - 1\right)} \cdot 2 \cdot 5 = 10.$$

The Puzzling Scribble

The drawing displayed in the puzzle cannot be drawn without lifting the pencil and without tracing any line more than once because the drawn graph *is not connected*, that is, from an arbitrary vertex of the graph, one cannot reach another arbitrary vertex by tracing along its edges.

The following statement is true for connected graphs:

> *A connected graph can be traced along a single path without visiting any edge more than once if the number of vertices of odd degree is 0 or 2.*

In graph theory, such a path is called an *Eulerian path*.

The statement can be proved by induction. First, we show the following: if a connected graph with n edges has no point of odd order, it can always be traced as required.

The case $n = 1$ cannot occur. The case $n = 2$ is shown in the illustration, and this graph can obviously be traced in the required way.

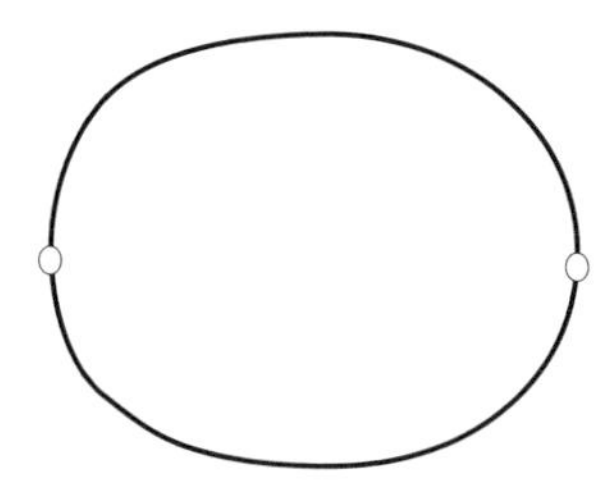

Assume that the statement is true for every connected graph with k edges for $k < n$ that has no vertex of odd degree (that is, assume such a graph can always be traced in the prescribed way).

Now we prove that the statement is true also for a graph with n edges, which is the induction step.

Consider an arbitrary vertex V on a graph with n edges. From V, we trace from one edge to the next (without visiting any edge twice) until we return to V. Sooner or later we must return there, since otherwise, the path would have to end in some other vertex, U. That cannot be the case, however. If we have crossed the vertex U before t times, then we have traced a total of $2t$ edges that end at U. Now, by way of a new edge, we have returned to U and can't leave again, meaning $2t + 1$ edges end at U. This contradicts the assumption that the degree of every vertex is even.

Thus, we have traced a complete round trip through V. Now we exclude every edge of this round trip and every vertex from which all the edges that extend from it have already been traced. If we have thus excluded all vertices, then we have traced the entire graph in the prescribed way. But if there are vertices still left, then it is true for the remaining graph that the degree of each of its vertices is even. Further, the remaining graph has fewer edges than the previous graph. We can't yet use the induction hypothesis, however, because it is not certain that the remaining graph is connected.

It can be demonstrated, though, that every connected portion can be traced separately. Further, we can see that because of the connectedness of the original graph, every remaining connected part touches the "circuit" that runs through V, which we had already excluded.

We can connect the two touching circuits as in the illustration.

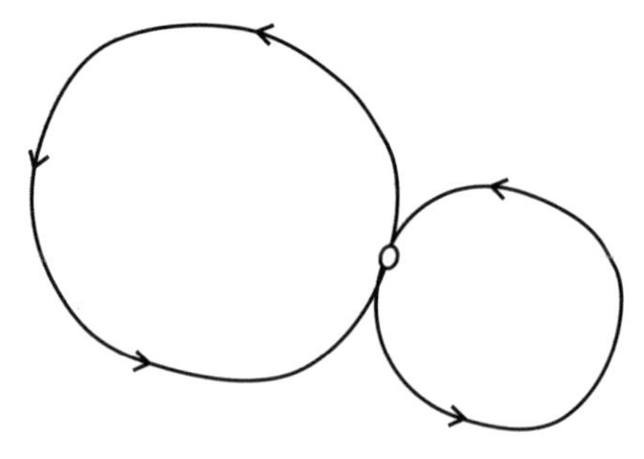

In other words, when making our original round trip from V, we take a detour and trace the complete circuit of the previously forgotten connected part. With this technique, we can trace the entire graph in the prescribed way. Thus, we have proved the statement.

Similarly, we see also the following: If, in the case of two vertices of odd degree, we start at such a vertex, we can only get stuck (reach a dead end) at the other such vertex. We now exclude the edges of a path so traced, as well as those points all of whose extending edges we had already traced. We then have connected graphs with smaller numbers of edges that no longer have any points of odd degree. These graphs can all be traced in the prescribed way and (as described above) can be connected to the path already traced and excluded.

It is also clear that otherwise (with vertices of odd degree not numbering 0 or 2) a graph cannot be traced in the prescribed manner (in keeping with the observation that "If we enter, we also have to exit").

49th Week

Footrace

The product of a runner's jersey number and his placement can be at most

$$12 \cdot 12 = 144.$$

We write down all those numbers up to 144 that, divided by 13, leave the remainder 1:

$$1,\ 14,\ 27,\ 40,\ 53,\ 66,\ 79,\ 92,\ 105,\ 118,\ 131,\ 144.$$

Of these, we exclude 53, 79, 92, 105, 118, and 131 because they cannot be factored into a product of two numbers both not greater than 12. We factor the remaining numbers in all possible ways such that both factors are at most 12:

$$1 = 1 \cdot 1, \quad 14 = 2 \cdot 7, \quad 27 = 3 \cdot 9, \quad 40 = 4 \cdot 10,$$
$$40 = 5 \cdot 8, \quad 66 = 6 \cdot 11, \quad 144 = 12 \cdot 12.$$

Since, except for 1 and 12 (numbers multiplied by themselves), all the numbers in the products above appear only once, the order in which the runners finished by jersey number was the following:

Place	1	2	3	4	5	6	7	8	9	10	11	12
Jersey number	1	7	9	10	8	11	2	5	3	4	6	12

No Tricks

We have to proceed from the circumstance that though Mr. Smith knows the number of the house, according to his statement the individual ages cannot be determined by the fact that their sum is equal to the number of the house. This means, in other words, that 1,296 has several representations as the product of three numbers such that the sum of the factors is the same.

So, we have to begin the solution of the puzzle by carrying out all the decompositions of 1,296 into the product of three numbers and then determining the sum of each set of factors. Since

$$1{,}296 = 2^4 \cdot 3^4,$$

the three factors in each decomposition have the forms

$$2^{a_1} \cdot 3^{b_1}, \quad 2^{a_2} \cdot 3^{b_2}, \quad 2^{a_3} \cdot 3^{b_3},$$

where $a_1 + a_2 + a_3 = b_1 + b_2 + b_3 = 4$. On this foundation, it's easy to represent all the decompositions of 1,296 into the product of three factors systematically. The sums of the three factors are shown in parentheses:

$1 \cdot 1 \cdot 1{,}296$	(1,298)	$1 \cdot 2 \cdot 648$	(651)	$1 \cdot 4 \cdot 324$	(329)
$1 \cdot 8 \cdot 162$	(171)	$1 \cdot 16 \cdot 81$	(98)	$1 \cdot 3 \cdot 432$	(436)
$1 \cdot 6 \cdot 216$	(223)	$1 \cdot 12 \cdot 108$	(121)	$1 \cdot 24 \cdot 54$	(79)
$1 \cdot 27 \cdot 48$	(76)	$1 \cdot 9 \cdot 144$	(154)	$1 \cdot 18 \cdot 72$	(91)
$1 \cdot 36 \cdot 36$	(73)	$2 \cdot 2 \cdot 324$	(328)	$2 \cdot 4 \cdot 162$	(168)
$2 \cdot 8 \cdot 81$	(91)	$2 \cdot 3 \cdot 216$	(221)	$2 \cdot 6 \cdot 108$	(116)
$2 \cdot 12 \cdot 54$	(68)	$2 \cdot 24 \cdot 27$	(53)	$2 \cdot 9 \cdot 72$	(83)
$2 \cdot 18 \cdot 36$	(56)	$4 \cdot 4 \cdot 81$	(89)	$4 \cdot 3 \cdot 108$	(115)
$4 \cdot 6 \cdot 54$	(64)	$4 \cdot 12 \cdot 27$	(43)	$4 \cdot 9 \cdot 36$	(49)
$4 \cdot 18 \cdot 18$	(40)	$8 \cdot 3 \cdot 54$	(65)	$8 \cdot 6 \cdot 27$	(41)
$8 \cdot 9 \cdot 18$	(35)	$16 \cdot 3 \cdot 27$	(46)	$16 \cdot 9 \cdot 9$	(34)
$3 \cdot 6 \cdot 72$	(81)	$3 \cdot 24 \cdot 18$	(45)	$3 \cdot 3 \cdot 144$	(150)
$3 \cdot 12 \cdot 36$	(51)	$3 \cdot 48 \cdot 9$	(60)	$6 \cdot 6 \cdot 36$	(48)
$6 \cdot 12 \cdot 18$	(36)	$6 \cdot 24 \cdot 9$	(39)	$12 \cdot 12 \cdot 9$	(33)

We see that the sum of the factors is the same in only two cases. So 91 is the number of the house, and the ages of the three residents are

$$1,\ 18,\ 72 \quad \text{or} \quad 2,\ 8,\ 81.$$

The old gentleman distinguishes the two solutions by explaining that all three are younger than he. So we can choose between the two solutions only if the old gentleman's age is between 72 and 81.

The ages of the three residents are thus 1, 18, and 72 years.

50th Week

The Balance Scale

Four weights suffice: a 1-pound weight, a 3-pound weight, a 9-pound weight, and a 27-pound weight.

In fact, with a little work we can see that for any k, the k initial powers of three—namely, $1,\ 3,\ 9,\ldots,\ 3^{k-1}$—will suffice to measure all weights from one all the way up to their sum—and each one in a unique manner!

We already know that any positive number n can be written in base 3, that is, in the form $\overline{a_m a_{m-1} \cdots a_1 a_0}$, where $n = a_m \cdot 3^m + \cdots + a_1 \cdot 3^1 + a_0 \cdot 1$ and each digit a_i is 0, 1, or 2. Less obvious but equally true is the fact that we can instead take each a_i to be 1, 0 or -1; for convenience, we replace -1 by the character #.

To see this, just take the rightmost 2 in a standard base-3 representation, turn it into a #, and then compensate by increasing the digit to its left by 1. The latter might become a 3, but that's all right because we can then change *it* to a 0 and bump up the next digit by one, etc. We do the same to the next 2 (moving right to left) and repeat until we have nothing left but 1s, 0s, and #s. For example,

$$25_{10} = 221_3 = 3\#1_3 = 10\#1_3 = 27_{10} - 3_{10} + 1_{10}.$$

This tells us that we can measure a 25-pound object by placing the 1-pound and 27-pound weights on one side of the beam scale and the object and the 3-pound weight on the other side.

Starting from any 1, 0, # number and reversing the above process shows that the representation is unique—there's only one way to arrange our power-of-3 weights on the scale to measure any given number up to $1 + 3 + 9 + 27 = 40$. Thus, our solution is efficient.

It follows that fewer than four weights will never do for our 1–40 range; in fact, three weights can measure no more than $1 + 3 + 9 = 13$ different values. With more work, one can show that, indeed, no set of four weights other than the one proposed can cover the range 1–40.

Unsuccessful Decree

Let $0 < x_1 \le x_2 \le \cdots \le x_{n-1} \le x_n$ be the monetary assets of each inhabitant. Since there was one richest man in the country, $x_{n-1} < x_n$. After doubling the assets of each person, the assets of no single inhabitant became greater than $2x_{n-1}$, and therefore, the equation

$$2x_{n-1} = x_n$$

must hold. By analogy, it must also hold that

$$2x_{i-1} = x_i \qquad (i = 2, \ldots, n);$$

that is, the assets of the individual inhabitants can be written as

$$x_{i+1} = x_1 \cdot 2^i \qquad (i = 1, 2, \ldots, n-1).$$

The distribution of assets can thus be carried out only according to this scheme. We still need to show, however, that such a distribution is possible.

After the doubling of everyone else's wealth, the remaining fortune of the originally richest man is

$$2^{n-1} \cdot x_1 - \left(x_1 + 2 \cdot x_1 + 2^2 \cdot x_1 + \cdots + 2^{n-3} \cdot x_1 + 2^{n-2} \cdot x_1\right).$$

Using the formula for the sum of a geometric series (see Section 3 of the Appendix, p. 223), we obtain the following for the fortune of the originally richest man:

$$2^{n-1} \cdot x_1 - \left(2^{n-1} \cdot x_1 - x_1\right) = x_1.$$

Thus, the distribution can, in fact, be carried out, and the fortune of the originally richest man becomes the fortune of the originally poorest man (under the conditions, there could be many).

51st Week

Book Lovers

We need to find prime numbers p and q that fulfill the equation:

$$p^a = q^b + 1,$$

where $a > 1$ and $b > 1$ are integers.

Both primes cannot be odd because then there would be an even number on one side of the above equation, and an odd number on the other. Thus, one of the two primes p and q is 2. Therefore, we have to find an odd prime r, whose nth power differs from 2^k by 1.

If n is odd, then

$$2^k = (r-1)\left(r^{n-1} + r^{n-2} + r^{n-3} + \cdots + r^2 + r + 1\right),$$

or

$$2^k = (r+1)\left(r^{n-1} - r^{n-2} + r^{n-3} - \cdots + r^2 - r + 1\right).$$

In both cases, the second factor on the right-hand side is greater than 1 and, since it is the sum of an odd number of odd integers, it is itself odd. But this is not possible since a divisor of 2^k cannot be odd.

So n is even; we can write it in the form $2m$, and the equations take the form

$$2^k = r^{2m} + 1 \tag{1}$$

or

$$2^k = r^{2m} - 1. \tag{2}$$

Equation (1) is not possible. There is a number divisible by 4 on the left-hand side, since $k > 1$; but on the right-hand side is a number which when divided by 4, yields the remainder 2, because it is a number greater by 1 than the square of an odd number.

Equation (2) can be written as follows:

$$2^k = (r^m)^2 - 1.$$

Using the fact that r is odd, we write $2x + 1$ for r^m:

$$2^k = (2x+1)^2 - 1 = 4x^2 + 4x.$$

Now we divide both sides of the equation by 2^2:

$$2^{k-2} = x(x+1).$$

This is possible only if $x = 1$, and therefore it follows that $r = 3$, $n = 2$, and $k = 3$.

Thus, Peter is on page eight and Thomas is on page nine.

A Party of Eight

By condition 3, Michael cannot be the judge.

Aside from Michael, every man's name begins with J. Therefore, using our first conclusion, the judge's name also begins with J, and given condition 5, his wife's name is Judith.

Given conditions 1 and 7, Josh and Joseph are siblings. By condition 6, neither one can be Judith's husband.

Judith's husband is Jacob, then, and Jacob is a judge by profession.

Margaret cannot be married. First, Jacob is already taken and Josh and Joseph are her siblings. Thus, only Michael could be Margaret's husband. That is, however, not possible; otherwise, Michael, Josh, and Joseph would be related. But among these three there must be two who are not related, since the physician and the bachelor, by condition 2, are not related.

Given the previous conclusion, then, Marsha is married, and by condition 4, Marsha's husband can only be the physician. Again by condition 4, her husband's name can only be Michael, since at least one of the two brothers Josh and Joseph is a tradesman.

By condition 8, Joseph is not a cabinetmaker. So, Joseph is the locksmith, and Josh is the cabinetmaker. Again by condition 8, we know that Joseph has a wife, and since Margaret is single, Maria is Joseph's wife.

In summary, Jacob is a judge (Judith is his wife), Michael is a physician (his wife is Marsha), Joseph is a locksmith (Maria is his wife), and Josh is a cabinetmaker (and a bachelor); Margaret is single.

52nd Week

Weighing Once Wins!

The clever engineer suggested the following plan:

"I take 1 ball bearing out of the first crate, 2 ball bearings out of the second crate, ..., 200 ball bearings out of the 200th crate. I put them all on a scale and weigh them. If all the ball bearings were of the new type, the total weight would be exactly

$$1 + 2 + 3 + \cdots + 200 = \frac{200 \cdot 201}{2} = 20{,}100 \text{ pounds.}$$

But the scale will register a deviation of n times 4 ounces, which means that the older type of ball bearings are in the nth crate. This deviation will be positive or negative, depending on whether the old ball bearings are heavier or lighter than the new ones."

Clever Heirs

Let x (expressed in dollars) be the whole estate, and let y be the amount an heir received.

The inheritance of the first heir was

$$y = a + \frac{x - a}{n}.$$

The inheritance of the second heir was

$$y = 2a + \frac{1}{n}\left(x - \left(a + \frac{x-a}{n}\right) - 2a\right).$$

From this it follows that $x = (n-1)^2 \cdot a$ and $y = (n-1) \cdot a$, from which we get $\frac{x}{y} = n - 1$ for the number of heirs.

We still need to check that according to this scheme each heir did in fact receive the same amount. The correctness of the statement can be proved by mathematical induction. We'll leave this proof to the reader.

Appendix

1. Terminology

We use very little terminology in this book. Since some of this terminology is used in textbooks with slightly different meanings or different names, to avoid confusion, let use define what we mean.

Numbers

An *integer* is one of the numbers

$$\ldots, -3,\ -2,\ -1,\ 0,\ 1,\ 2,\ 3,\ \ldots$$

Many books call these *whole numbers*. So 27 and -15 are integers, 2.7 is not.

From 1 on, the integers are called *natural numbers*. So 27 and 101 are natural numbers, 0, -4, and 2.7 are not.

If we also include 0, we get the *nonnegative integers*:

$$0,\ 1,\ 2,\ 3,\ \ldots$$

As you see, the natural numbers are the positive integers.

Operations

When we add numbers, they are referred to as *summands*. When we multiply them, they are called *factors*. We denote multiplication by a centered dot: $23 \cdot 5$, except if we multiply a number with a symbol, for instance, $25x$, we drop the dot.

In a division, the number you divide is the *dividend*, the number you divide with is the *divisor*.

If you divide a natural number, n with the natural number d, you always get

$$n = d \cdot k + r,$$

where k is the result of the division, and r is between 0 and $d - 1$. This integer r is called the *remainder*.

Basic concepts

We define many important concepts in the text, here are some:

arithmetic and geometric sequences p. 224
arithmetic and geometric mean p. 91
binary system p. 108
binomial coefficient p. 164
decimal system p. 107
directed graph p. 187
factorial p. 100
Fibonacci sequence p. 139
graph p. 168
Pascal's triangle p. 163
prime number p. 73
quadratic equation p. 224

2. Mathematical Induction

Mathematical induction is a method of proof by which it is possible to prove a statement about the elements of series by first proving its correctness for the first element, and then showing that by assuming the correctness of the statement *for an element* of the series, the correctness of the statement *for the succeeding element* also follows, that is, *the statement carries*.

The following statement can, for example, be proven by induction:

> For the sum of the squares of the first n integers, the following equation holds:
>
> $$1^2 + 2^2 + 3^2 + \cdots + n^2 = \frac{n(n+1)(2n+1)}{6}.$$

The proof by induction consists of two parts:

1. *Induction base:* We determine that the equation is correct for $n = 1$:

$$\frac{1 \cdot 2 \cdot 3}{6} = 1.$$

2. *Induction step:* Let k be any integer. We assume that the statement is correct for k and then prove that the correctness of the statement for k carries to $k+1$. The induction hypothesis is

$$1^2 + 2^2 + \cdots + k^2 = \frac{k \cdot (k+1) \cdot (2k+1)}{6}. \tag{1}$$

We want to prove that equation (1) holds also for $k+1$, that is,

$$1^2 + 2^2 + \cdots + k^2 + (k+1)^2 = \frac{(k+1) \cdot (k+2) \cdot (2k+3)}{6}.$$

We add $(k+1)^2$ to both sides of equation (1):

$$1^2 + 2^2 + \cdots + k^2 + (k+1)^2 = \frac{k \cdot (k+1) \cdot (2k+1)}{6} + (k+1)^2.$$

The right-hand side can be rearranged as follows:

$$\frac{k \cdot (k+1) \cdot (2k+1) + 6(k+1)^2}{6} = \frac{(k+1) \cdot (k+2) \cdot (2k+3)}{6}.$$

This means that

$$1^2 + 2^2 + \cdots + (k+1)^2 = \frac{(k+1) \cdot (k+2) \cdot (2k+3)}{6}.$$

That is exactly what we wished to prove.

3. Some Important Formulas

Here are some standard expansions:

$$\begin{aligned}
(a+b)^2 &= a^2 + ab + ab + b^2 = a^2 + 2ab + b^2, \\
(a-b)^2 &= a^2 - 2ab + b^2, \\
(a+b)^3 &= a^3 + 3a^2b + 3ab^2 + b^3, \\
(a-b)^3 &= a^3 - 3a^2b + 3ab^2 - b^3, \\
(a+b)(a-b) &= a^2 - b^2.
\end{aligned}$$

For *every integer* n, it is true that

$$a^n - b^n = (a-b)(a^{n-1} + a^{n-2}b + a^{n-3}b^2 + \cdots + b^{n-1}).$$

In particular, for $n = 3$ and for $n = 4$, it is true that

$$a^3 - b^3 = (a-b)(a^2 + ab + b^2),$$
$$a^4 - b^4 = (a-b)(a^3 + a^2b + ab^2 + b^3).$$

For any *even* n, it is true that

$$a^n - b^n = (a+b)\left(a^{n-1} - a^{n-2}b + a^{n-3}b^2 - + \cdots - b^{n-1}\right).$$

For any *odd* n, it is true that

$$a^n + b^n = (a+b)\left(a^{n-1} - a^{n-2}b + a^{n-3}b^2 + \cdots + b^{n-1}\right).$$

In particular, for $n = 3$ and for $n = 4$, it is true that

$$a^3 + b^3 = (a+b)\left(a^2 - ab + b^2\right),$$
$$a^4 - b^4 = (a+b)\left(a^3 - a^2b + ab^2 - b^3\right).$$

An *arithmetic sequence* is a sequence of numbers in which (from its second member) the *difference* (d) between any member and the immediately preceding member is constant.

The nth member of the series is $a_n = a_1 + (n-1)d$.

Let S_n be the sum of the first n members of an arithmetic series. This sum can be calculated as follows:

$$S_n = n \cdot \frac{a_1 + a_n}{2}.$$

A *geometric sequence* is a sequence of numbers in which (from its second member) the quotient of a member (as numerator) and the immediately preceding member (as denominator) is a constant q.

The nth member of the sequence is $a_n = a_1 \cdot q^{n-1}$.

Let G_n be the sum of the first n members of a geometric sequence. This sum can be calculated as follows:

$$G_n = a_1 \cdot \frac{q^n - 1}{q - 1} \quad (q \neq 1).$$

The *quadratic equation* $ax^2 + bx + c = 0$ $(a \neq 0)$ has the following two solutions:

$$x_1 = \frac{-b + \sqrt{b^2 - 4ac}}{2a},$$

$$x_2 = \frac{-b - \sqrt{b^2 - 4ac}}{2a},$$

where x_1 and x_2 are real numbers, and it holds that $b^2 - 4ac \geq 0$.

4. Prime Numbers

2	263	607	983	1381	1783	2221	2663	3083	3541	4001
3	269	613	991	1399	1787	2237	2671	3089	3547	4003
5	271	617	997	1409	1789	2239	2677	3109	3557	4007
7	277	619	1009	1423	1801	2243	2683	3119	3559	4013
11	281	631	1013	1427	1811	2251	2687	3121	3571	4019
13	283	641	1019	1429	1823	2267	2689	3137	3581	4021
17	293	643	1021	1433	1831	2269	2693	3163	3583	4027
19	307	647	1031	1439	1847	2273	2699	3167	3593	4049
23	311	653	1033	1447	1861	2281	2707	3169	3607	4051
29	313	659	1039	1451	1867	2287	2711	3181	3613	4057
31	317	661	1049	1453	1871	2293	2713	3187	3617	4073
37	331	673	1051	1459	1873	2297	2719	3191	3623	4079
41	337	677	1061	1471	1877	2309	2729	3203	3631	4091
43	347	683	1063	1481	1879	2311	2731	3209	3637	4093
47	349	691	1069	1483	1889	2333	2741	3217	3643	4099
53	353	701	1087	1487	1901	2339	2749	3221	3659	4111
59	359	709	1091	1489	1907	2341	2753	3229	3671	4127
61	367	719	1093	1493	1913	2347	2767	3251	3673	4129
67	373	727	1097	1499	1931	2351	2777	3253	3677	4133
71	379	733	1103	1511	1933	2357	2789	3257	3691	4139
73	383	739	1109	1523	1949	2371	2791	3259	3697	4153
79	389	743	1117	1531	1951	2377	2797	3271	3701	4157
83	397	751	1123	1543	1973	2381	2801	3299	3709	4159
89	401	757	1129	1549	1979	2383	2803	3301	3719	4177
97	409	761	1151	1553	1987	2389	2819	3307	3727	4201
101	419	769	1153	1559	1993	2393	2833	3313	3733	4211
103	421	773	1163	1567	1997	2399	2837	3319	3739	4217
107	431	787	1171	1571	1999	2411	2843	3323	3761	4219
109	433	797	1181	1579	2003	2417	2851	3329	3767	4229
113	439	809	1187	1583	2011	2423	2857	3331	3769	4231
127	443	811	1193	1597	2017	2437	2861	3343	3779	4241
131	449	821	1201	1601	2027	2441	2879	3347	3793	4243
137	457	823	1213	1607	2029	2447	2887	3359	3797	4253
139	461	827	1217	1609	2039	2459	2897	3361	3803	4259
149	463	829	1223	1613	2053	2467	2903	3371	3821	4261
151	467	839	1229	1619	2063	2473	2909	3373	3823	4271
157	479	853	1231	1621	2069	2477	2917	3389	3833	4273
163	487	857	1237	1627	2081	2503	2927	3391	3847	4283
167	491	859	1249	1637	2083	2521	2939	3407	3851	4289
173	499	863	1259	1657	2087	2531	2953	3413	3853	4297
179	503	877	1277	1663	2089	2539	2957	3433	3863	4327
181	509	881	1279	1667	2099	2543	2963	3449	3877	4337
191	521	883	1283	1669	2111	2549	2969	3457	3881	4339
193	523	887	1289	1693	2113	2551	2971	3461	3889	4349
197	541	907	1291	1697	2129	2557	2999	3463	3907	4357
199	547	911	1297	1699	2131	2579	3001	3467	3911	4363
211	557	919	1301	1709	2137	2591	3011	3469	3917	4373
223	563	929	1303	1721	2141	2593	3019	3491	3919	4391
227	569	937	1307	1723	2143	2609	3023	3499	3923	4397
229	571	941	1319	1733	2153	2617	3037	3511	3929	4409
233	577	947	1321	1741	2161	2621	3041	3517	3931	4421
239	587	953	1327	1747	2179	2633	3049	3527	3943	4423
241	593	967	1361	1753	2203	2647	3061	3529	3947	4441
251	599	971	1367	1759	2207	2657	3067	3533	3967	4447
257	601	977	1373	1777	2213	2659	3079	3539	3989	4451

5. Prime Factorization

$4 = 2^2$
$6 = 2 \cdot 3$
$8 = 2^3$

$9 = 3^2$
$10 = 2 \cdot 5$
$12 = 2^2 \cdot 3$

$14 = 2 \cdot 7$
$15 = 3 \cdot 5$
$16 = 2^4$

$18 = 2 \cdot 3^2$
$20 = 2^2 \cdot 5$
$21 = 3 \cdot 7$

$22 = 2 \cdot 11$
$24 = 2^3 \cdot 3$
$25 = 5^2$

$26 = 2 \cdot 13$
$27 = 3^3$
$28 = 2^2 \cdot 7$

$30 = 2 \cdot 3 \cdot 5$
$32 = 2^5$
$33 = 3 \cdot 11$

$34 = 2 \cdot 17$
$35 = 5 \cdot 7$
$36 = 2^2 \cdot 3^2$

$38 = 2 \cdot 19$
$39 = 3 \cdot 13$
$40 = 2^3 \cdot 5$

$42 = 2 \cdot 3 \cdot 7$
$44 = 2^2 \cdot 11$
$45 = 3^2 \cdot 5$

$46 = 2 \cdot 23$
$48 = 2^4 \cdot 3$
$49 = 7^2$

$50 = 2 \cdot 5^2$
$51 = 3 \cdot 17$
$52 = 2^2 \cdot 13$

$54 = 2 \cdot 3^3$
$55 = 5 \cdot 11$
$56 = 2^3 \cdot 7$

$57 = 3 \cdot 19$
$58 = 2 \cdot 29$
$60 = 2^2 \cdot 3 \cdot 5$

$62 = 2 \cdot 31$
$63 = 3^2 \cdot 7$
$64 = 2^6$

$65 = 5 \cdot 13$
$66 = 2 \cdot 3 \cdot 11$
$68 = 2^2 \cdot 17$

$69 = 3 \cdot 23$
$70 = 2 \cdot 5 \cdot 7$
$72 = 2^3 \cdot 3^2$

$74 = 2 \cdot 37$
$75 = 3 \cdot 5^2$
$76 = 2^2 \cdot 19$

$77 = 7 \cdot 11$
$78 = 2 \cdot 3 \cdot 13$
$80 = 2^4 \cdot 5$

$81 = 3^4$
$82 = 2 \cdot 41$
$84 = 2^2 \cdot 3 \cdot 7$

$85 = 5 \cdot 17$
$86 = 2 \cdot 43$
$87 = 3 \cdot 29$

$88 = 2^3 \cdot 11$
$90 = 2 \cdot 3^2 \cdot 5$
$91 = 7 \cdot 13$

$92 = 2^2 \cdot 23$
$93 = 3 \cdot 31$
$94 = 2 \cdot 47$

$95 = 5 \cdot 19$
$96 = 2^5 \cdot 3$
$98 = 2 \cdot 7^2$

$99 = 3^2 \cdot 11$
$100 = 2^2 \cdot 5^2$
$102 = 2 \cdot 3 \cdot 17$

$104 = 2^3 \cdot 13$
$105 = 3 \cdot 5 \cdot 7$
$106 = 2 \cdot 53$

$108 = 2^2 \cdot 3^3$
$110 = 2 \cdot 5 \cdot 11$
$111 = 3 \cdot 37$

$112 = 2^4 \cdot 7$
$114 = 2 \cdot 3 \cdot 19$
$115 = 5 \cdot 23$

$116 = 2^2 \cdot 29$
$117 = 3^2 \cdot 13$
$118 = 2 \cdot 59$

$119 = 7 \cdot 17$
$120 = 2^3 \cdot 3 \cdot 5$
$121 = 11^2$

$122 = 2 \cdot 61$
$123 = 3 \cdot 41$
$124 = 2^2 \cdot 31$

$125 = 5^3$
$126 = 2 \cdot 3^2 \cdot 7$
$128 = 2^7$

$129 = 3 \cdot 43$
$130 = 2 \cdot 5 \cdot 13$
$132 = 2^2 \cdot 3 \cdot 11$

$133 = 7 \cdot 19$
$134 = 2 \cdot 67$
$135 = 3^3 \cdot 5$

$136 = 2^3 \cdot 17$
$138 = 2 \cdot 3 \cdot 23$
$140 = 2^2 \cdot 5 \cdot 7$

$141 = 3 \cdot 47$
$142 = 2 \cdot 71$
$143 = 11 \cdot 13$

$144 = 2^4 \cdot 3^2$
$145 = 5 \cdot 29$
$146 = 2 \cdot 73$

$147 = 3 \cdot 7^2$
$148 = 2^2 \cdot 37$
$150 = 2 \cdot 3 \cdot 5^2$

$152 = 2^3 \cdot 19$
$153 = 3^2 \cdot 17$
$154 = 2 \cdot 7 \cdot 11$

$155 = 5 \cdot 31$
$156 = 2^2 \cdot 3 \cdot 13$
$158 = 2 \cdot 79$

$159 = 3 \cdot 53$
$160 = 2^5 \cdot 5$
$161 = 7 \cdot 23$

$162 = 2 \cdot 3^4$
$164 = 2^2 \cdot 41$
$165 = 3 \cdot 5 \cdot 11$

$166 = 2 \cdot 83$
$168 = 2^3 \cdot 3 \cdot 7$
$169 = 13^2$

$170 = 2 \cdot 5 \cdot 17$
$171 = 3^2 \cdot 19$
$172 = 2^2 \cdot 43$

$174 = 2 \cdot 3 \cdot 29$
$175 = 5^2 \cdot 7$
$176 = 2^4 \cdot 11$

$177 = 3 \cdot 59$
$178 = 2 \cdot 89$
$180 = 2^2 \cdot 3^2 \cdot 5$

$182 = 2 \cdot 7 \cdot 13$
$183 = 3 \cdot 61$
$184 = 2^3 \cdot 23$

$185 = 5 \cdot 37$
$186 = 2 \cdot 3 \cdot 31$
$187 = 11 \cdot 17$

$188 = 2^2 \cdot 47$
$189 = 3^3 \cdot 7$
$190 = 2 \cdot 5 \cdot 19$

$192 = 2^6 \cdot 3$
$194 = 2 \cdot 97$
$195 = 3 \cdot 5 \cdot 13$

$196 = 2^2 \cdot 7^2$
$198 = 2 \cdot 3^2 \cdot 11$
$200 = 2^3 \cdot 5^2$

$201 = 3 \cdot 67$
$202 = 2 \cdot 101$
$203 = 7 \cdot 29$

$204 = 2^2 \cdot 3 \cdot 17$
$205 = 5 \cdot 41$
$206 = 2 \cdot 103$

$207 = 3^2 \cdot 23$
$208 = 2^4 \cdot 13$
$209 = 11 \cdot 19$

$210 = 2 \cdot 3 \cdot 5 \cdot 7$
$212 = 2^2 \cdot 53$
$213 = 3 \cdot 71$

$214 = 2 \cdot 107$
$215 = 5 \cdot 43$
$216 = 2^3 \cdot 3^3$

$217 = 7 \cdot 31$
$218 = 2 \cdot 109$
$219 = 3 \cdot 73$

$220 = 2^2 \cdot 5 \cdot 11$
$221 = 13 \cdot 17$
$222 = 2 \cdot 3 \cdot 37$

$224 = 2^5 \cdot 7$
$225 = 3^2 \cdot 5^2$
$226 = 2 \cdot 113$

$228 = 2^2 \cdot 3 \cdot 19$
$230 = 2 \cdot 5 \cdot 23$
$231 = 3 \cdot 7 \cdot 11$

$232 = 2^3 \cdot 29$
$234 = 2 \cdot 3^2 \cdot 13$
$235 = 5 \cdot 47$

$236 = 2^2 \cdot 59$
$237 = 3 \cdot 79$
$238 = 2 \cdot 7 \cdot 17$

$240 = 2^4 \cdot 3 \cdot 5$
$242 = 2 \cdot 11^2$
$243 = 3^5$

$244 = 2^2 \cdot 61$
$245 = 5 \cdot 7^2$
$246 = 2 \cdot 3 \cdot 41$

$247 = 13 \cdot 19$
$248 = 2^3 \cdot 31$
$249 = 3 \cdot 83$

$250 = 2 \cdot 5^3$
$252 = 2^2 \cdot 3^2 \cdot 7$
$253 = 11 \cdot 23$

$254 = 2 \cdot 127$
$255 = 3 \cdot 5 \cdot 17$
$256 = 2^8$

$258 = 2 \cdot 3 \cdot 43$
$259 = 7 \cdot 37$
$260 = 2^2 \cdot 5 \cdot 13$

$261 = 3^2 \cdot 29$
$262 = 2 \cdot 131$
$264 = 2^3 \cdot 3 \cdot 11$

$265 = 5 \cdot 53$
$266 = 2 \cdot 7 \cdot 19$
$267 = 3 \cdot 89$

$268 = 2^2 \cdot 67$
$270 = 2 \cdot 3^3 \cdot 5$
$272 = 2^4 \cdot 17$

$273 = 3 \cdot 7 \cdot 13$
$274 = 2 \cdot 137$
$275 = 5^2 \cdot 11$

$276 = 2^2 \cdot 3 \cdot 23$
$278 = 2 \cdot 139$
$279 = 3^2 \cdot 31$

$280 = 2^3 \cdot 5 \cdot 7$
$282 = 2 \cdot 3 \cdot 47$
$284 = 2^2 \cdot 71$

$285 = 3 \cdot 5 \cdot 19$
$286 = 2 \cdot 11 \cdot 13$
$287 = 7 \cdot 41$

$288 = 2^5 \cdot 3^2$
$289 = 17^2$
$290 = 2 \cdot 5 \cdot 29$

$291 = 3 \cdot 97$
$292 = 2^2 \cdot 73$
$294 = 2 \cdot 3 \cdot 7^2$

$295 = 5 \cdot 59$
$296 = 2^3 \cdot 37$
$297 = 3^3 \cdot 11$

$298 = 2 \cdot 149$
$299 = 13 \cdot 23$
$300 = 2^2 \cdot 3 \cdot 5^2$

$301 = 7 \cdot 43$
$302 = 2 \cdot 151$
$303 = 3 \cdot 101$

$304 = 2^4 \cdot 19$
$305 = 5 \cdot 61$
$306 = 2 \cdot 3^2 \cdot 17$

$308 = 2^2 \cdot 7 \cdot 11$
$309 = 3 \cdot 103$
$310 = 2 \cdot 5 \cdot 31$

$312 = 2^3 \cdot 3 \cdot 13$
$314 = 2 \cdot 157$
$315 = 3^2 \cdot 5 \cdot 7$

$316 = 2^2 \cdot 79$
$318 = 2 \cdot 3 \cdot 53$
$319 = 11 \cdot 29$

$320 = 2^6 \cdot 5$
$321 = 3 \cdot 107$
$322 = 2 \cdot 7 \cdot 23$

$323 = 17 \cdot 19$
$324 = 2^2 \cdot 3^4$
$325 = 5^2 \cdot 13$

$326 = 2 \cdot 163$
$327 = 3 \cdot 109$
$328 = 2^3 \cdot 41$

$329 = 7 \cdot 47$
$330 = 2 \cdot 3 \cdot 5 \cdot 11$
$332 = 2^2 \cdot 83$

$333 = 3^2 \cdot 37$
$334 = 2 \cdot 167$
$335 = 5 \cdot 67$

$336 = 2^4 \cdot 3 \cdot 7$
$338 = 2 \cdot 13^2$
$339 = 3 \cdot 113$

$340 = 2^2 \cdot 5 \cdot 17$
$341 = 11 \cdot 31$
$342 = 2 \cdot 3^2 \cdot 19$

$343 = 7^3$
$344 = 2^3 \cdot 43$
$345 = 3 \cdot 5 \cdot 23$

$346 = 2 \cdot 173$
$348 = 2^2 \cdot 3 \cdot 29$
$350 = 2 \cdot 5^2 \cdot 7$

$351 = 3^3 \cdot 13$
$352 = 2^5 \cdot 11$
$354 = 2 \cdot 3 \cdot 59$

$355 = 5 \cdot 71$
$356 = 2^2 \cdot 89$
$357 = 3 \cdot 7 \cdot 17$

$358 = 2 \cdot 179$
$360 = 2^3 \cdot 3^2 \cdot 5$
$361 = 19^2$

$362 = 2 \cdot 181$
$363 = 3 \cdot 11^2$
$364 = 2^2 \cdot 7 \cdot 13$

$365 = 5 \cdot 73$
$366 = 2 \cdot 3 \cdot 61$
$368 = 2^4 \cdot 23$

$369 = 3^2 \cdot 41$
$370 = 2 \cdot 5 \cdot 37$
$371 = 7 \cdot 53$

$372 = 2^2 \cdot 3 \cdot 31$
$374 = 2 \cdot 11 \cdot 17$
$375 = 3 \cdot 5^3$

$376 = 2^3 \cdot 47$
$377 = 13 \cdot 29$
$378 = 2 \cdot 3^3 \cdot 7$

$380 = 2^2 \cdot 5 \cdot 19$
$381 = 3 \cdot 127$
$382 = 2 \cdot 191$

$384 = 2^7 \cdot 3$
$385 = 5 \cdot 7 \cdot 11$
$386 = 2 \cdot 193$

$387 = 3^2 \cdot 43$
$388 = 2^2 \cdot 97$
$390 = 2 \cdot 3 \cdot 5 \cdot 13$

$391 = 17 \cdot 23$
$392 = 2^3 \cdot 7^2$
$393 = 3 \cdot 131$

$394 = 2 \cdot 197$
$395 = 5 \cdot 79$
$396 = 2^2 \cdot 3^2 \cdot 11$

$398 = 2 \cdot 199$
$399 = 3 \cdot 7 \cdot 19$
$400 = 2^4 \cdot 5^2$

$402 = 2 \cdot 3 \cdot 67$
$403 = 13 \cdot 31$
$404 = 2^2 \cdot 101$

$405 = 3^4 \cdot 5$
$406 = 2 \cdot 7 \cdot 29$
$407 = 11 \cdot 37$

$408 = 2^3 \cdot 3 \cdot 17$
$410 = 2 \cdot 5 \cdot 41$
$411 = 3 \cdot 137$

$412 = 2^2 \cdot 103$
$413 = 7 \cdot 59$
$414 = 2 \cdot 3^2 \cdot 23$

$415 = 5 \cdot 83$
$416 = 2^5 \cdot 13$
$417 = 3 \cdot 139$

$418 = 2 \cdot 11 \cdot 19$
$420 = 2^2 \cdot 3 \cdot 5 \cdot 7$
$422 = 2 \cdot 211$

$423 = 3^2 \cdot 47$
$424 = 2^3 \cdot 53$
$425 = 5^2 \cdot 17$

$426 = 2 \cdot 3 \cdot 71$
$427 = 7 \cdot 61$
$428 = 2^2 \cdot 107$

$429 = 3 \cdot 11 \cdot 13$
$430 = 2 \cdot 5 \cdot 43$
$432 = 2^4 \cdot 3^3$

$434 = 2 \cdot 7 \cdot 31$
$435 = 3 \cdot 5 \cdot 29$
$436 = 2^2 \cdot 109$

$437 = 19 \cdot 23$
$438 = 2 \cdot 3 \cdot 73$
$440 = 2^3 \cdot 5 \cdot 11$

$441 = 3^2 \cdot 7^2$
$442 = 2 \cdot 13 \cdot 17$
$444 = 2^2 \cdot 3 \cdot 37$

$445 = 5 \cdot 89$
$446 = 2 \cdot 223$
$447 = 3 \cdot 149$

$448 = 2^6 \cdot 7$
$450 = 2 \cdot 3^2 \cdot 5^2$
$451 = 11 \cdot 41$

$452 = 2^2 \cdot 113$
$453 = 3 \cdot 151$
$454 = 2 \cdot 227$

$455 = 5 \cdot 7 \cdot 13$
$456 = 2^3 \cdot 3 \cdot 19$
$458 = 2 \cdot 229$

$459 = 3^3 \cdot 17$
$460 = 2^2 \cdot 5 \cdot 23$
$462 = 2 \cdot 3 \cdot 7 \cdot 11$

$464 = 2^4 \cdot 29$
$465 = 3 \cdot 5 \cdot 31$
$466 = 2 \cdot 233$

$468 = 2^2 \cdot 3^2 \cdot 13$
$469 = 7 \cdot 67$
$470 = 2 \cdot 5 \cdot 47$

$471 = 3 \cdot 157$
$472 = 2^3 \cdot 59$
$473 = 11 \cdot 43$

$474 = 2 \cdot 3 \cdot 79$
$475 = 5^2 \cdot 19$
$476 = 2^2 \cdot 7 \cdot 17$

$477 = 3^2 \cdot 53$
$478 = 2 \cdot 239$
$480 = 2^5 \cdot 3 \cdot 5$

$481 = 13 \cdot 37$
$482 = 2 \cdot 241$
$483 = 3 \cdot 7 \cdot 23$

$484 = 2^2 \cdot 11^2$
$485 = 5 \cdot 97$
$486 = 2 \cdot 3^5$

$488 = 2^3 \cdot 61$
$489 = 3 \cdot 163$
$490 = 2 \cdot 5 \cdot 7^2$

$492 = 2^2 \cdot 3 \cdot 41$
$493 = 17 \cdot 29$
$494 = 2 \cdot 13 \cdot 19$

$495 = 3^2 \cdot 5 \cdot 11$
$496 = 2^4 \cdot 31$
$497 = 7 \cdot 71$

$498 = 2 \cdot 3 \cdot 83$
$500 = 2^2 \cdot 5^3$
$501 = 3 \cdot 167$

$502 = 2 \cdot 251$
$504 = 2^3 \cdot 3^2 \cdot 7$
$505 = 5 \cdot 101$

$506 = 2 \cdot 11 \cdot 23$
$507 = 3 \cdot 13^2$
$508 = 2^2 \cdot 127$

$510 = 2 \cdot 3 \cdot 5 \cdot 17$
$511 = 7 \cdot 73$
$512 = 2^9$

$513 = 3^3 \cdot 19$
$514 = 2 \cdot 257$
$515 = 5 \cdot 103$

$516 = 2^2 \cdot 3 \cdot 43$
$517 = 11 \cdot 47$
$518 = 2 \cdot 7 \cdot 37$

$519 = 3 \cdot 173$
$520 = 2^3 \cdot 5 \cdot 13$
$522 = 2 \cdot 3^2 \cdot 29$

$524 = 2^2 \cdot 131$
$525 = 3 \cdot 5^2 \cdot 7$
$526 = 2 \cdot 263$

$527 = 17 \cdot 31$
$528 = 2^4 \cdot 3 \cdot 11$
$529 = 23^2$

$530 = 2 \cdot 5 \cdot 53$
$531 = 3^2 \cdot 59$
$532 = 2^2 \cdot 7 \cdot 19$

$533 = 13 \cdot 41$
$534 = 2 \cdot 3 \cdot 89$
$535 = 5 \cdot 107$

$536 = 2^3 \cdot 67$
$537 = 3 \cdot 179$
$538 = 2 \cdot 269$

$539 = 7^2 \cdot 11$
$540 = 2^2 \cdot 3^3 \cdot 5$
$542 = 2 \cdot 271$

$543 = 3 \cdot 181$
$544 = 2^5 \cdot 17$
$545 = 5 \cdot 109$

$546 = 2 \cdot 3 \cdot 7 \cdot 13$
$548 = 2^2 \cdot 137$
$549 = 3^2 \cdot 61$

$550 = 2 \cdot 5^2 \cdot 11$
$551 = 19 \cdot 29$
$552 = 2^3 \cdot 3 \cdot 23$

$553 = 7 \cdot 79$
$554 = 2 \cdot 277$
$555 = 3 \cdot 5 \cdot 37$

$556 = 2^2 \cdot 139$
$558 = 2 \cdot 3^2 \cdot 31$
$559 = 13 \cdot 43$

$560 = 2^4 \cdot 5 \cdot 7$
$561 = 3 \cdot 11 \cdot 17$
$562 = 2 \cdot 281$

$564 = 2^2 \cdot 3 \cdot 47$
$565 = 5 \cdot 113$
$566 = 2 \cdot 283$

$567 = 3^4 \cdot 7$
$568 = 2^3 \cdot 71$
$570 = 2 \cdot 3 \cdot 5 \cdot 19$

$572 = 2^2 \cdot 11 \cdot 13$
$573 = 3 \cdot 191$
$574 = 2 \cdot 7 \cdot 41$

$575 = 5^2 \cdot 23$
$576 = 2^6 \cdot 3^2$
$578 = 2 \cdot 17^2$

$579 = 3 \cdot 193$
$580 = 2^2 \cdot 5 \cdot 29$
$581 = 7 \cdot 83$

$582 = 2 \cdot 3 \cdot 97$
$583 = 11 \cdot 53$
$584 = 2^3 \cdot 73$

$585 = 3^2 \cdot 5 \cdot 13$
$586 = 2 \cdot 293$
$588 = 2^2 \cdot 3 \cdot 7^2$

$589 = 19 \cdot 31$
$590 = 2 \cdot 5 \cdot 59$
$591 = 3 \cdot 197$

$592 = 2^4 \cdot 37$
$594 = 2 \cdot 3^3 \cdot 11$
$595 = 5 \cdot 7 \cdot 17$

$596 = 2^2 \cdot 149$
$597 = 3 \cdot 199$
$598 = 2 \cdot 13 \cdot 23$

$600 = 2^3 \cdot 3 \cdot 5^2$
$602 = 2 \cdot 7 \cdot 43$
$603 = 3^2 \cdot 67$

$604 = 2^2 \cdot 151$
$605 = 5 \cdot 11^2$
$606 = 2 \cdot 3 \cdot 101$

$608 = 2^5 \cdot 19$
$609 = 3 \cdot 7 \cdot 29$
$610 = 2 \cdot 5 \cdot 61$

$611 = 13 \cdot 47$
$612 = 2^2 \cdot 3^2 \cdot 17$
$614 = 2 \cdot 307$

$615 = 3 \cdot 5 \cdot 41$
$616 = 2^3 \cdot 7 \cdot 11$
$618 = 2 \cdot 3 \cdot 103$

$620 = 2^2 \cdot 5 \cdot 31$
$621 = 3^3 \cdot 23$
$622 = 2 \cdot 311$

$623 = 7 \cdot 89$
$624 = 2^4 \cdot 3 \cdot 13$
$625 = 5^4$

$626 = 2 \cdot 313$
$627 = 3 \cdot 11 \cdot 19$
$628 = 2^2 \cdot 157$

$629 = 17 \cdot 37$
$630 = 2 \cdot 3^2 \cdot 5 \cdot 7$
$632 = 2^3 \cdot 79$

$633 = 3 \cdot 211$
$634 = 2 \cdot 317$
$635 = 5 \cdot 127$

$636 = 2^2 \cdot 3 \cdot 53$
$637 = 7^2 \cdot 13$
$638 = 2 \cdot 11 \cdot 29$

$639 = 3^2 \cdot 71$
$640 = 2^7 \cdot 5$
$642 = 2 \cdot 3 \cdot 107$

$644 = 2^2 \cdot 7 \cdot 23$
$645 = 3 \cdot 5 \cdot 43$
$646 = 2 \cdot 17 \cdot 19$

$648 = 2^3 \cdot 3^4$
$649 = 11 \cdot 59$
$650 = 2 \cdot 5^2 \cdot 13$

$651 = 3 \cdot 7 \cdot 31$
$652 = 2^2 \cdot 163$
$654 = 2 \cdot 3 \cdot 109$

$655 = 5 \cdot 131$
$656 = 2^4 \cdot 41$
$657 = 3^2 \cdot 73$

$658 = 2 \cdot 7 \cdot 47$
$660 = 2^2 \cdot 3 \cdot 5 \cdot 11$
$662 = 2 \cdot 331$

$663 = 3 \cdot 13 \cdot 17$
$664 = 2^3 \cdot 83$
$665 = 5 \cdot 7 \cdot 19$

$666 = 2 \cdot 3^2 \cdot 37$
$667 = 23 \cdot 29$
$668 = 2^2 \cdot 167$

$669 = 3 \cdot 223$
$670 = 2 \cdot 5 \cdot 67$
$671 = 11 \cdot 61$

$672 = 2^5 \cdot 3 \cdot 7$
$674 = 2 \cdot 337$
$675 = 3^3 \cdot 5^2$

$676 = 2^2 \cdot 13^2$
$678 = 2 \cdot 3 \cdot 113$
$679 = 7 \cdot 97$

$680 = 2^3 \cdot 5 \cdot 17$
$681 = 3 \cdot 227$
$682 = 2 \cdot 11 \cdot 31$

$684 = 2^2 \cdot 3^2 \cdot 19$
$685 = 5 \cdot 137$
$686 = 2 \cdot 7^3$

$687 = 3 \cdot 229$
$688 = 2^4 \cdot 43$
$689 = 13 \cdot 53$

$690 = 2 \cdot 3 \cdot 5 \cdot 23$
$692 = 2^2 \cdot 173$
$693 = 3^2 \cdot 7 \cdot 11$

$694 = 2 \cdot 347$
$695 = 5 \cdot 139$
$696 = 2^3 \cdot 3 \cdot 29$

$697 = 17 \cdot 41$
$698 = 2 \cdot 349$
$699 = 3 \cdot 233$

$700 = 2^2 \cdot 5^2 \cdot 7$
$702 = 2 \cdot 3^3 \cdot 13$
$703 = 19 \cdot 37$

$704 = 2^6 \cdot 11$
$705 = 3 \cdot 5 \cdot 47$
$706 = 2 \cdot 353$

$707 = 7 \cdot 101$
$708 = 2^2 \cdot 3 \cdot 59$
$710 = 2 \cdot 5 \cdot 71$

$711 = 3^2 \cdot 79$
$712 = 2^3 \cdot 89$
$713 = 23 \cdot 31$

$714 = 2 \cdot 3 \cdot 7 \cdot 17$
$715 = 5 \cdot 11 \cdot 13$
$716 = 2^2 \cdot 179$

$717 = 3 \cdot 239$
$718 = 2 \cdot 359$
$720 = 2^4 \cdot 3^2 \cdot 5$

$721 = 7 \cdot 103$
$722 = 2 \cdot 19^2$
$723 = 3 \cdot 241$

$724 = 2^2 \cdot 181$
$725 = 5^2 \cdot 29$
$726 = 2 \cdot 3 \cdot 11^2$

$728 = 2^3 \cdot 7 \cdot 13$
$729 = 3^6$
$730 = 2 \cdot 5 \cdot 73$

$731 = 17 \cdot 43$
$732 = 2^2 \cdot 3 \cdot 61$
$734 = 2 \cdot 367$

$735 = 3 \cdot 5 \cdot 7^2$
$736 = 2^5 \cdot 23$
$737 = 11 \cdot 67$

$738 = 2 \cdot 3^2 \cdot 41$
$740 = 2^2 \cdot 5 \cdot 37$
$741 = 3 \cdot 13 \cdot 19$

$742 = 2 \cdot 7 \cdot 53$
$744 = 2^3 \cdot 3 \cdot 31$
$745 = 5 \cdot 149$

$746 = 2 \cdot 373$
$747 = 3^2 \cdot 83$
$748 = 2^2 \cdot 11 \cdot 17$

$749 = 7 \cdot 107$
$750 = 2 \cdot 3 \cdot 5^3$
$752 = 2^4 \cdot 47$

$753 = 3 \cdot 251$
$754 = 2 \cdot 13 \cdot 29$
$755 = 5 \cdot 151$

$756 = 2^2 \cdot 3^3 \cdot 7$
$758 = 2 \cdot 379$
$759 = 3 \cdot 11 \cdot 23$

$760 = 2^3 \cdot 5 \cdot 19$
$762 = 2 \cdot 3 \cdot 127$
$763 = 7 \cdot 109$

$764 = 2^2 \cdot 191$
$765 = 3^2 \cdot 5 \cdot 17$
$766 = 2 \cdot 383$

$767 = 13 \cdot 59$
$768 = 2^8 \cdot 3$
$770 = 2 \cdot 5 \cdot 7 \cdot 11$

$771 = 3 \cdot 257$
$772 = 2^2 \cdot 193$
$774 = 2 \cdot 3^2 \cdot 43$

$775 = 5^2 \cdot 31$
$776 = 2^3 \cdot 97$
$777 = 3 \cdot 7 \cdot 37$

$778 = 2 \cdot 389$
$779 = 19 \cdot 41$
$780 = 2^2 \cdot 3 \cdot 5 \cdot 13$

$781 = 11 \cdot 71$
$782 = 2 \cdot 17 \cdot 23$
$783 = 3^3 \cdot 29$

$784 = 2^4 \cdot 7^2$
$785 = 5 \cdot 157$
$786 = 2 \cdot 3 \cdot 131$

$788 = 2^2 \cdot 197$
$789 = 3 \cdot 263$
$790 = 2 \cdot 5 \cdot 79$

$791 = 7 \cdot 113$
$792 = 2^3 \cdot 3^2 \cdot 11$
$793 = 13 \cdot 61$

$794 = 2 \cdot 397$
$795 = 3 \cdot 5 \cdot 53$
$796 = 2^2 \cdot 199$

$798 = 2 \cdot 3 \cdot 7 \cdot 19$
$799 = 17 \cdot 47$
$800 = 2^5 \cdot 5^2$

$801 = 3^2 \cdot 89$
$802 = 2 \cdot 401$
$803 = 11 \cdot 73$

$804 = 2^2 \cdot 3 \cdot 67$
$805 = 5 \cdot 7 \cdot 23$
$806 = 2 \cdot 13 \cdot 31$

$807 = 3 \cdot 269$
$808 = 2^3 \cdot 101$
$810 = 2 \cdot 3^4 \cdot 5$

$812 = 2^2 \cdot 7 \cdot 29$
$813 = 3 \cdot 271$
$814 = 2 \cdot 11 \cdot 37$

$815 = 5 \cdot 163$
$816 = 2^4 \cdot 3 \cdot 17$
$817 = 19 \cdot 43$

$818 = 2 \cdot 409$
$819 = 3^2 \cdot 7 \cdot 13$
$820 = 2^2 \cdot 5 \cdot 41$

$822 = 2 \cdot 3 \cdot 137$
$824 = 2^3 \cdot 103$
$825 = 3 \cdot 5^2 \cdot 11$

$826 = 2 \cdot 7 \cdot 59$
$828 = 2^2 \cdot 3^2 \cdot 23$
$830 = 2 \cdot 5 \cdot 83$

$831 = 3 \cdot 277$
$832 = 2^6 \cdot 13$
$833 = 7^2 \cdot 17$

$834 = 2 \cdot 3 \cdot 139$
$835 = 5 \cdot 167$
$836 = 2^2 \cdot 11 \cdot 19$

$837 = 3^3 \cdot 31$
$838 = 2 \cdot 419$
$840 = 2^3 \cdot 3 \cdot 5 \cdot 7$

$841 = 29^2$
$842 = 2 \cdot 421$
$843 = 3 \cdot 281$

$844 = 2^2 \cdot 211$
$845 = 5 \cdot 13^2$
$846 = 2 \cdot 3^2 \cdot 47$

$847 = 7 \cdot 11^2$
$848 = 2^4 \cdot 53$
$849 = 3 \cdot 283$

$850 = 2 \cdot 5^2 \cdot 17$
$851 = 23 \cdot 37$
$852 = 2^2 \cdot 3 \cdot 71$

$854 = 2 \cdot 7 \cdot 61$
$855 = 3^2 \cdot 5 \cdot 19$
$856 = 2^3 \cdot 107$

$858 = 2 \cdot 3 \cdot 11 \cdot 13$
$860 = 2^2 \cdot 5 \cdot 43$
$861 = 3 \cdot 7 \cdot 41$

$862 = 2 \cdot 431$
$864 = 2^5 \cdot 3^3$
$865 = 5 \cdot 173$

$866 = 2 \cdot 433$
$867 = 3 \cdot 17^2$
$868 = 2^2 \cdot 7 \cdot 31$

$869 = 11 \cdot 79$
$870 = 2 \cdot 3 \cdot 5 \cdot 29$
$871 = 13 \cdot 67$

$872 = 2^3 \cdot 109$
$873 = 3^2 \cdot 97$
$874 = 2 \cdot 19 \cdot 23$

$875 = 5^3 \cdot 7$
$876 = 2^2 \cdot 3 \cdot 73$
$878 = 2 \cdot 439$

$879 = 3 \cdot 293$
$880 = 2^4 \cdot 5 \cdot 11$
$882 = 2 \cdot 3^2 \cdot 7^2$

$884 = 2^2 \cdot 13 \cdot 17$
$885 = 3 \cdot 5 \cdot 59$
$886 = 2 \cdot 443$

$888 = 2^3 \cdot 3 \cdot 37$
$889 = 7 \cdot 127$
$890 = 2 \cdot 5 \cdot 89$

$891 = 3^4 \cdot 11$
$892 = 2^2 \cdot 223$
$893 = 19 \cdot 47$

$894 = 2 \cdot 3 \cdot 149$
$895 = 5 \cdot 179$
$896 = 2^7 \cdot 7$

$897 = 3 \cdot 13 \cdot 23$
$898 = 2 \cdot 449$
$899 = 29 \cdot 31$

$900 = 2^2 \cdot 3^2 \cdot 5^2$
$901 = 17 \cdot 53$
$902 = 2 \cdot 11 \cdot 41$

$903 = 3 \cdot 7 \cdot 43$
$904 = 2^3 \cdot 113$
$905 = 5 \cdot 181$

$906 = 2 \cdot 3 \cdot 151$
$908 = 2^2 \cdot 227$
$909 = 3^2 \cdot 101$

$910 = 2 \cdot 5 \cdot 7 \cdot 13$
$912 = 2^4 \cdot 3 \cdot 19$
$913 = 11 \cdot 83$

$914 = 2 \cdot 457$
$915 = 3 \cdot 5 \cdot 61$
$916 = 2^2 \cdot 229$

$917 = 7 \cdot 131$
$918 = 2 \cdot 3^3 \cdot 17$
$920 = 2^3 \cdot 5 \cdot 23$

$921 = 3 \cdot 307$
$922 = 2 \cdot 461$
$923 = 13 \cdot 71$

$924 = 2^2 \cdot 3 \cdot 7 \cdot 11$
$925 = 5^2 \cdot 37$
$926 = 2 \cdot 463$

$927 = 3^2 \cdot 103$
$928 = 2^5 \cdot 29$
$930 = 2 \cdot 3 \cdot 5 \cdot 31$

$931 = 7^2 \cdot 19$
$932 = 2^2 \cdot 233$
$933 = 3 \cdot 311$

$934 = 2 \cdot 467$
$935 = 5 \cdot 11 \cdot 17$
$936 = 2^3 \cdot 3^2 \cdot 13$

$938 = 2 \cdot 7 \cdot 67$
$939 = 3 \cdot 313$
$940 = 2^2 \cdot 5 \cdot 47$

$942 = 2 \cdot 3 \cdot 157$
$943 = 23 \cdot 41$
$944 = 2^4 \cdot 59$

$945 = 3^3 \cdot 5 \cdot 7$
$946 = 2 \cdot 11 \cdot 43$
$948 = 2^2 \cdot 3 \cdot 79$

$949 = 13 \cdot 73$
$950 = 2 \cdot 5^2 \cdot 19$
$951 = 3 \cdot 317$

$952 = 2^3 \cdot 7 \cdot 17$
$954 = 2 \cdot 3^2 \cdot 53$
$955 = 5 \cdot 191$

$956 = 2^2 \cdot 239$
$957 = 3 \cdot 11 \cdot 29$
$958 = 2 \cdot 479$

$959 = 7 \cdot 137$
$960 = 2^6 \cdot 3 \cdot 5$
$961 = 31^2$

$962 = 2 \cdot 13 \cdot 37$
$963 = 3^2 \cdot 107$
$964 = 2^2 \cdot 241$

$965 = 5 \cdot 193$
$966 = 2 \cdot 3 \cdot 7 \cdot 23$
$968 = 2^3 \cdot 11^2$

$969 = 3 \cdot 17 \cdot 19$
$970 = 2 \cdot 5 \cdot 97$
$972 = 2^2 \cdot 3^5$

$973 = 7 \cdot 139$
$974 = 2 \cdot 487$
$975 = 3 \cdot 5^2 \cdot 13$

$976 = 2^4 \cdot 61$
$978 = 2 \cdot 3 \cdot 163$
$979 = 11 \cdot 89$

$980 = 2^2 \cdot 5 \cdot 7^2$
$981 = 3^2 \cdot 109$
$982 = 2 \cdot 491$

$984 = 2^3 \cdot 3 \cdot 41$
$985 = 5 \cdot 197$
$986 = 2 \cdot 17 \cdot 29$

$987 = 3 \cdot 7 \cdot 47$
$988 = 2^2 \cdot 13 \cdot 19$
$989 = 23 \cdot 43$

$990 = 2 \cdot 3^2 \cdot 5 \cdot 11$
$992 = 2^5 \cdot 31$
$993 = 3 \cdot 331$

$994 = 2 \cdot 7 \cdot 71$
$995 = 5 \cdot 199$
$996 = 2^2 \cdot 3 \cdot 83$

$998 = 2 \cdot 499$
$999 = 3^3 \cdot 37$
$1000 = 2^3 \cdot 5^3$

$1001 = 7 \cdot 11 \cdot 13$
$1002 = 2 \cdot 3 \cdot 167$
$1003 = 17 \cdot 59$

$1004 = 2^2 \cdot 251$
$1005 = 3 \cdot 5 \cdot 67$
$1006 = 2 \cdot 503$

$1007 = 19 \cdot 53$
$1008 = 2^4 \cdot 3^2 \cdot 7$
$1010 = 2 \cdot 5 \cdot 101$

A Friendly Afterword to the Reader

I turn to you, now, dear reader. Just now, having solved every problem, you meant to put the book down, thinking you've finished it. I direct myself also to those curious readers who may just have thumbed through the pages to the end, without having diligently solved all the problems. Or could it be you've not even started solving them?

Let me say a few words about why, how, and from what sources this book was written.

In retrospect, be it said this book belongs to the large family of books meant to popularize mathematics. Its primary goal, as set forth in the foreword, is calisthenics for the brain. To reach this goal, though, I've chosen particular puzzles not so much for their own sake; rather, I've picked brain teasers that acquaint the reader with mathematics, and even make the experience pleasurable.

Given the book's design, the scope of mathematics it teaches is rather modest, but that is only fitting, for the reader must discover the theorems presented here for himself. In several of the brain teasers, interesting mathematical theorems are reduced to a series of elementary steps leading to the result. Think, for example, of the series of problems leading to the theory of Nim games, or of the puzzles that lead to the Eulerian condition for the traversability of graphs. A discovery like that is a terrific experience! After one's first such experience, mathematics is no longer just a collection of theorems created by some god (or by humans, but who knows how?). We realize that we too can create theorems!

Given that this math book is a work of popular science, it follows, of course, that much of its content is not new. "Nothing new under the sun" is an adage applicable also to popular math—or at least rarely does something new appear. But those readers at whom this book is aimed presumably won't be exhaustively versed in the world literature of popular mathematics. So, for them, it's all the same if a brain teaser here is based on a puzzle that has appeared elsewhere or is even a familiar classic. It may well be that a classic brain teaser is all the better for having stood the test of time and assumed a corresponding aura.

My collection of brain teasers relies heavily on the beautiful book *Mathematical Recreations* by M. Kraitchik, (London, George Allen & Unwill Ltd, 1955). I've also borrowed a number of problems from the Hungarian *Mathematical and Physical Journal for Secondary Schools.*[9]

Finally, let me note that in putting this book together, I've also used brain teasers that are not puzzles in the strict sense of the word. In those cases, however, some mathematical goal has always been in the background.

Thanks for your attention, and once again, have fun!

George Grätzer

---

[9]The Hungarian title is *Középiskolai Matematikai és Fizikai Lapok.*

Afterword to the English Translation

I encountered the young man in Winnipeg at the Polo Park shopping center.[10] He was, of course, much faster than I was, so I saw only his back as he walked past me. I thought he looked familiar. He was thin, had thick brown hair, and was in a big hurry. I caught up with him as he stopped in front of a store. He turned half way towards me and knew at once who I was. I can't say he was overjoyed to see me.

"Things haven't gone all that badly," I said, in a somewhat defensive fashion.

"Really?" he replied. "You're the son of the 'Puzzle King,'[11] and as a junior at university, you wrote a book which met your family's expectations: your first puzzle book. So, how many puzzle books have you added to the family tradition in the half century since then?"

"Not a single one. But I've written 20 other books. I've become a researcher in the field of mathematics. The two most important books I've written were on my two main research fields, and both have appeared in several editions. I've published more than 230 research papers. And I've become an expert in mathematical typesetting. The

[10]To a Hungarian reader it would be clear that I am following the story line of the *Young Man* by Frigyes Karinthy(1887–1938), a Hungarian author of pithy short stories and plays.

[11]József Grätzer (1897–1945), the Hungarian Puzzle King, wrote in the late 1930s (among other titles) the puzzle books *Rébusz* and *Sicc*, which have brought pleasure to many generations of young readers.

books I've written on this subject have found their way into the hands of more than 40, 000 mathematicians."

"Good," said the young man, "I know you've worked on a lot of different things. But why haven't you pursued puzzles?"

"I didn't just quit puzzles. When I wrote this book, I was one of the leading Hungarian chess composers[12] and I won many competitions.[13] I quit doing that too. Thinking up puzzles, creating chess problems, and mathematical research—all three processes run partly consciously and partly subconsciously. We ponder a problem for a long time, and one morning we wake up with the solution. It's possible to do our conscious thinking in two or three different areas. It's much more difficult, though, to do the same subconsciously."

The young man didn't agree. I didn't pursue my conversation with him. Perhaps he was right, after all.

* * *

This book is the translation of the second Hungarian edition of this puzzle book. How did this second edition come about after more than 50 years?

My father's book, *Sicc*, has been published some 10 times since the war. The contract with the last publisher lapsed in 2007 and I was approached by Ádám Halmos, another Hungarian publisher, to republish *Sicc* along with *Rébusz*.

In her last year of high school, the publisher's mother, Mária Halmos, won a mathematical contest for which she was rewarded with a copy of the first edition of my puzzle book. In the course of her career, she worked with talented high school math students and she has been using my book to get them motivated.

At the meeting to discuss the publication of my father's two books, I proposed publishing a *Grätzer Trilogy*: my father's two books and my own puzzle book. With Mária present, I had an instant ally. And who can say "no" to his mother? Thus, the trilogy was published.

For the second edition, two experts have scrutinized each puzzle thoroughly: Mária Halmos, whom we've already met, and Erika Kuczmann, who for two decades pursued the teaching of mathematics to talented high school students. These two women have refined nearly every puzzle and solution and corrected many errors. It has been my great good fortune to have had these two prominent experts so fruitfully revise my book.

[12]A *chess composer* is a person who creates endgame studies or other chess problems.

[13]See http://www.magyarsakkszerzok.com/ and http://www.magyarsakkszerzok.com/gratzer-gy.htm.

The book was first translated into German and from German to English. I would like to thank my publisher, Klaus Peters, and my editor, Charlotte Henderson, for their commitment to quality books. I was also assisted by Karen Kipper. Mária Halmos contributed a great deal to this project; her attention to detail is legendary.

It gives me great pleasure to know my readers can now take the product of these efforts in hand.

George Grätzer